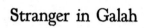

Stranger in Galah

Stranger in Galah

MICHAEL BARRETT

W · W · NORTON & COMPANY · INC ·
NEW YORK

To
FRANCES

I

DEANE lifted bleary eyes from the track. He frowned, crouching over the driving-wheel, concentrated hazily . . .

Dust stirred on the horizon, cloudy, smoking. There was movement in that great silence, burning under the sun. A trail of red sand, rising across the scrub. It was too small for another vehicle: an animal or a man, alone, a speck in the vastness, travelling.

The Austin truck rode on, jarring upon its lashed-up springs, crashing down into every pitfall of the unmade road. Deane steered, watched the other track of red dust journey over the bush as a larger red cloud founted into the harsh dry air behind him, marking his own passage.

Open sky, hard white light: nothing in the sky and nothing else on the land. Red dust and dry vegetation, parched yellow spinifex, barren mulga scrub. Away on the edge of the sky came low and flat-topped hills. The way stretched ahead, laced through the red sand.

The other track of dust came nearer; and soon the thing at its head was two-legged, human, running fast. Running desperately . . . Nearer still and it was a man, dark-skinned against the yellow-white landscape, black against red earth. An aborigine.

Deane stared behind; but there was nothing else to be

7

seen through the settling dust. The man rushed on. The Austin rolled forward down the track.

Their paths were due to cross.

The heat had got Deane, burning, choking. At the back of the vehicle, liquid splashed in the spare water cans, the petrol cans, under the canvas awning. Dust-dry land and the drought was on, perishing.

The road was long in front and long behind, over gravel and rock outcrop, over saltpan and baked creek, the longest road. And Deane had kept going because he had found nowhere to stop.

He rubbed the back of his hand over his dry lips. His eyes were red-rimmed; he swung the wheel against the ridges, coughing that red dust in his throat.

The aborigine was racing, straining, throwing himself madly across the rough ground in the shimmering, mirage-making heat. But he was real enough: he had an open shirt, full black hair on a head which lolled, lurched from side to side as he went. He was wild and urgent, thick lips drawn back, white teeth flashing in a jagged, fearful grin.

For the first time he seemed to see the truck. He swerved, jerked out of his frantic concentration. His gaping mouth twisted; he flung himself to one side, away from the track, went leaping across the bush in fiercer, more desperate frenzy.

A hundred yards away from Deane he fell. He crashed down, out of sight among the spinifex grass. He came up to his feet, rushed on. Then he sprawled, dropped full-length again. His head struggled, then he collapsed. The yellow burnt spinifex hid him.

8

Deane waited, his foot pausing on the accelerator. He pressed down, then released. At last, reluctantly, he braked. He switched off the engine.

Silence, full and complete. The dust settled; Deane got down from the truck. The sun and silence hammered down from a wide blue sky. He took one more look around the emptiness. He swore briefly. Then he reached for his water-bottle, hitched up his trousers and walked with the same slow reluctance away from the truck and towards the spot where the aborigine had plunged out of sight.

The spinifex brushed at his loose trousers. Deane stood still, staring down. The aborigine rolled over and gazed at him, panting; he cowered like an exhausted animal awaiting the kill, shaking his black shoulders, his chest heaving. Sweat covered the glistening skin: the man was small, thin-bodied, with waves of thick black hair, a strongly-sculpted face which was torn by pain and terror. He wore ripped khaki shirt and dirty dungarees, and he was bare-footed. His black eyes lifted to Deane's.

Deane said, 'What the hell's *your* bloody hurry?'

He looked back over the bush, and it remained empty as a dead land. No breath of wind, but silence.

He dropped on his knees, offered the water. The aborigine made a dry noise in his throat; he turned his head away and struggled to rise. For a moment the man's eyes searched over the quiet, sunstruck plain, and then a fresh fear flared over him, dilating his eyes. He fell back to the red sandy earth. His chest arched rigidly.

Deane was hot and weary, burdened with plenty

9

trouble enough among his own affairs. He stoppered-up the water can, raked his mind irritably for fragments of the childish abo-English he had picked up on the slow journey north.

At last he said, 'You-feller, why you bin runnin'? What's-a-matter, longa you boy?'

He bent over the aborigine and again the man squirmed away. They were alone together in the emptiness, in the silence: no animal, no bird, the hot air holding the echoes of the aborigine's panting breath. Beads of sweat started out on his flat and ugly face with its squashed black nose: his eyes were elemental, live and tormented.

Deane said again, 'You tell 'im. You tell 'im, heh?' His voice was harsher in its reluctance, snapping out the pidgin words.

The aborigine's head rolled again, and he pressed himself away into the earth.

'You tell 'im. You tell 'im.' Deane gripped the man's shoulder.

The aborigine yelped and pulled himself back once more: he tried to escape but he was finished. He fell down among the rustling spinifex. Again the black, fearful eyes lifted to Deane.

Deane watched them with a kind of superior scorn, mixed with curiosity. He raised his own eyes to stare over the bush. Now at last he could see further movement. The dust was rising again, spurting red against the blue horizon and the distant hills. Motion, towards them.

He waited. The trail of dust threaded a swift course; light glinted on something bright at its head. Silence lay still over the bush.

He glanced back at the aborigine. In their very nearness, their waiting, something linked them. The aborigine did not move again. His chest heaved, but he was quiet. His bare feet and thin ankles protruded from the tattered dungaree trouser-ends. Once again he met Deane's gaze, trembling.

Suddenly he said, 'Boss . . . Aw, boss . . .' His voice was thick with terror. He turned to look at the approaching dust-trail.

Deane stirred at last. They were both men, in their different ways. Something touched him, like humanity; rare and unusual feeling. He said, 'What's it all about?'

The aborigine's teeth chattered, but he did not speak.

Now the bright thing at the head of the dust-cloud was the windscreen of a truck. The vehicle was a Land-Rover: it jolted through the scrub, drawing nearer.

Deane and the aborigine watched it come, across the dusty earth in the glaring light, remote, approaching. The air was dead. Noise of the engine rose. Four men were inside the truck; two were black and two were white.

The Land-Rover drove past Deane's Austin, off the track, over to them. Deane got to his feet. The Land-Rover pulled up among the spinifex and the driver switched off. Silence swelled up again.

The man at the wheel was a fleshy-jawed white man in a grey coat and broad-brimmed Ashburton hat. The aborigine stockman next to him wore garish cowboy jacket: he grinned triumphantly as he looked at the man beside Deane on the ground. In the truck another scared-faced aborigine sat with a young white man who carried a rifle across his knee.

11

Deane stood still. At last he lifted his hand in gesture of casual salute. The heavy white man got down from the driving seat and strode across. He stared at the aborigine on the ground, then at Deane.

His gaze probed Deane from cold eyes. He had a big body and a hard face: his clothes were of good cloth. A gold pin fixed his neckerchief to his shirt. The wide hatbrim shadowed his eyes and the heavy lips were set straight and unsmiling.

He said, 'Who are you?'

Deane paused. 'The name's Deane.'

'Where are you heading for?'

Deane shrugged. 'Where the road goes.'

The big man turned, and the sun caught his eyes. They were fixed on Deane, considering. Then his gaze swivelled away: he stared at the man on the ground, and the aborigine looked back at him dumbly.

The white man said, 'We've been looking for this feller.'

'He doesn't seem so blooming keen to see you.'

The big man grunted. Then he said, 'Gettup.'

The aborigine shivered, but he did not move.

The big man called, 'Here, Ken.'

The aborigine stockman jumped out of the Land-Rover.

'Put him in the back.'

'Yes, boss.' The red-braided jacket flashed in the sun as the stockman bent, jerked the man from the ground to his feet. He pushed him across to the Land-Rover; he kicked at him and shoved him into the rear beside the other silent aborigine. The younger white man jerked the rifle muzzle, gesturing the newcomer into place.

12

The boss white man stared briefly at Deane. 'Had some trouble with my abos.' Then he turned on his heel, went back to the vehicle.

Deane glanced across. For a moment he caught the gaze of the man they had come to collect. The aborigine's face was bare now, frozen of feeling. His eyes met Deane's as he sat in the back of the truck. They looked at one another.

The black eyes were as empty as the black face. Something flickered in them for an instant, an emotion which would stir and break. Entreaty, pleading. Then the flicker died. The aborigine stared away.

Deane stepped aside. He half-moved, then stopped.

The Land-Rover's wheels spun in the soft earth. It gripped, reversing. It swung round, accelerated down the track into the direction from which it had come. The ploughing wheels flung up dust and soil: red, choking grit drifted over Deane where he stood.

He went back to the Austin. He took a drink of luke-warm water, sat gazing through the windscreen for a while. Then he started up the lurching engine, let in the drive, heaved on slowly down the empty road. The tyre marks of the Land-Rover unrolled themselves in front.

Heat and silence pressed over him. The unsprung chassis thudded into the hollows, almost breaking its back with every fall. Deane watched the speedometer; eighty miles from the last homestead, about twenty to the next.

He found himself remembering the dumb look on the face of the aborigine . . . He shook himself impatiently, changing his position on the warm leather of the seat.

The scrub grew thicker. Mulga trees dotted the bush. Ahead, a line of wire-fencing stretched for miles across the countryside. The track ran up to it: there was a gate and a crudely-painted sign: *Shut This Flaming Gate.*

Deane left the engine rattling to itself; he walked over and opened the gate. He wondered if this were the station from which the Land-Rover had come. He drove into the paddock and went on down the trail. No sign of homestead, of town, of habitation. Spinifex and scrub, saltbush, red sand and dry earth. Drought, sun and sky, the splashing of water and petrol behind him.

He passed his hand over an unshaven chin. The cracked rear-view mirror held a distorted convex view of the interior of the cab: the trail which unwound behind, the dusty parcels, tins and bags which were the swag he carried with him, west from Sydney, north even to Darwin and the sea if he should get that far. And on the seat the harsh-faced monkey, the unshaven ape, hairy-armed, red-eyed: the shiralee, the real burden he carried with him all the way: himself. Because he knew he was no bloody good.

He was too dry to spit. Twenty miles, and maybe there'd be a cool night, a beer, a sleep. A place to rest. And if he found nowhere to stay, he'd go on.

The paddock was dry and barren ground. The mulga trees came thicker, yet most of them were dead. Stiff, stark and mournful shapes, skeletal branches held out like green-grey bones. The Austin passed beside them. The track ran straight: ahead, Deane saw something strange, a dark shape among the grey arms of the dead trees near the road.

14

The truck went more slowly as he approached the spot. At last he stopped and got out. He stood in the beating sun and still air. An aborigine was hanging from the branches of one of the mulga trees. There was a rope around his neck and he was newly dead.

He wore khaki shirt and dungaree trousers with bare calloused feet poking through, and it was the man whose flight had crossed with Deane's on the plain.

Deane stood looking at him. Then he stared over the empty countryside. At last he gripped the still-swaying feet and tried to release the body. But the rope was firmly tied. Then he went over to the trunk of the tree, swung his weight against it.

The dead wood was ant-eaten and hollow right through. The tree swayed, and then it split and fell like a giant toppling, splintering, crashing. The dry noise of breaking echoed. The body of the aborigine came down with it, tumbling slackly to the earth.

Deane bent over the corpse among the broken sticks of wood. But there was nothing for him to do.

He stared around him again, at the dry land, then up at the sky. He saw two kites, high up in the blue.

He hesitated; he was going to pile brushwood over the limp black figure with its hideous face, drive on. None of his cursed business . . . He saw again the last, hopeless look on the face of the aborigine. Stupid pity moved him, from somewhere deep down.

Suddenly he changed his mind. He swore harshly again; then he lifted the thin body, carried it to the truck and put it in the back. He got into the seat and drove on.

The trail continued, the dust-cloud rose, the water splashed and the body lolled on the metal floor behind him. Heat spread under the cab roof as the Austin came out into the open from the mulga trees. A few miles on, Deane drove past a fork in the rough road, and then he reached an edge of the paddock, the infinite stretches of fencing wire again.

He went through the gate, down the track. The ground broke from its flatness, passed into low hills, rock outcrops and dry creeks. Five miles on, the way wound round a hillock: the land beyond opened up into a plain and he saw the small town he had waited for.

An empty road, a scattering of single-storeyed shacks, caught in white sunlight. The Austin truck rattled on loose stones and then Deane drove it down, bearing the aborigine's body: driving carefully, like a man bringing in a dangerous load, through the silent main street of Galah.

2

THE township was empty, dusty and ugly. On the edge came the overgrown cemetery, sand-covered white stones; and then the buildings, square and derelict, with pointed roofs and iron catchment tanks, motionless windmill generators.

Deane drove the truck by. A café with white lettering daubed on the walls, a general store: buckets and brooms, hurricane lanterns hung on hooks outside.

Cyclones had passed here, the willie-willies; houses were tumbled down, and grass grew between the broken walls, in the gaps of vacant plots. The town had the air of falling, of dying into desolation. There was a large, locked courthouse with shuttered and silted windows, where a dog sat in the shade of a parched bottle-tree. Late afternoon heat was heavy, pouring over the open street.

Deane's careful eyes probed behind the windscreen of the Austin as he went. He had come a long way, but in each new town the bitter caution stayed with him . . .

The hotel was a weatherboard and galvanized-iron structure with a covered veranda and a placard. *Galah Creek Hotel.* Nobody in sight but a couple of loungers who squatted behind the rails with their hats pushed back, their legs stretched out in front of them. They watched him.

A few children played in the street, white and black. There were lacey eucalyptus gums, silver and salmon-coloured, blue beyond the dry creek. Farther out, sunlight reflected upon the tin dwellings and wurlies of the aborigines, set in wasteland.

Deane reached the police station. He pulled to a halt: the lashed-up springs gave finally as his temporary repair of mulga stakes snapped. The chassis collapsed heavily on the wheels. Deane muttered. He got out, stood in the burning sun. For a moment he eyed the station, the corrugated-iron box with the heat waves shimmering high above it.

Stiffly, he straightened his back and walked round to the rear of the truck. The children had followed him

17

down the street. They were staring in, at the body where already the greedy flies had begun to collect; they were chattering, pointing.

Deane said, 'Clear off. Get out of it.'

He flung a canvas sheet over the dead aborigine. The group of young children split up and they ran away down the street, scattering, screaming out excitedly.

Deane hitched his trousers, turned and inspected the police station once more. The heat of the sun fell on his back. 'Hell,' he said. 'Why should I stick my sodding neck out . . . ?'

The abo was cold dead. He was only running himself into further chance of trouble. He caught just one more glimpse of those silent, stricken black eyes which had looked out across the bush at him. And he remembered fiercely all the small retreats and cowardices of the past which had led him here, still fleeing. Something stuck in him at last, thick, heavy bile.

He kicked back viciously and desperately; his boot-tip in the dry earth sent stones and grit showering, pattering on the station wooden floor.

Then slowly, he walked inside . . .

The office was sweltering hot; air roasted under the iron roof. A policeman in khaki uniform shirt sat at the desk, smoking, writing carefully. He looked up at Deane, nodded.

'G'day.'

'Afternoon,' Deane said.

'You from Kooni? Heard there was a truck coming through.'

'Yes.'

18

'Where are you heading for?' The policeman put down his pipe.

'North.' He spoke indifferently.

'You're way off the bitumen.'

'I know.' Deane watched him with the antagonism he had for all policemen. But the man looked reasonable; not young, hair beginning to grey, unaggressive and stolid face. Deane added, 'I picked something up, few miles outside town.'

'What?' The policeman lifted thick eyebrows.

Deane said, 'I found a dead abo hanging from a mulga tree. So I pulled him down and brought him in.'

The policeman's expression changed abruptly. A shadow passed behind his eyes. He asked sharply, 'Hanging?'

'Yes . . .' Deane stared around the bare office, then he turned back to the policeman. 'I caught up with him before that, further out in the bush. He was running fast, and mighty scared. Then a couple of johnnies in a Land-Rover drove along and collected him. When I followed down the road through the paddock I found him again. They'd strung him up high.' His voice was dispassionate.

The policeman stood, moved round behind his desk. He smoothed back his grey hair with one hand, and he seemed disturbed. Then he said, 'Where's the body?'

Deane jerked his head. 'Out in the truck.'

They went outside, into the harsh light. Deane pulled off the sheet. The flies swept up.

The policeman peered into the back.

Deane said, 'Know him?'

The policeman straightened up, replacing the sheet. He glanced across the open street. Now there were three or four men outside the pub-hotel, tall, dusty men, long-legged. An old truck was parked in the shade. A couple of other men on the veranda were staring over to the police station. A plump-bodied, middle-aged woman had come to the door of the general store and she too was watching the Austin and the police station. The children had collected near the café wall. An aborigine in a dirty red shirt squatted in the dust. The town was waking to a show of life.

Deane repeated, 'Know him?'

The policeman hesitated. At last he said, 'Yes, I know him.'

'Who is he?'

'One of the blacks from a station near here.'

'Which station is that?'

He paused again. 'Clancy Rock.'

Deane said, 'That's the one with the paddock back along the road? Where I found him?'

Once more the policeman hesitated. 'Yes.'

'They told me I'd pass through it when I left Kooni. Run by a bloke named Malone?'

'Yes.'

'Maybe I've met Mr. Malone,' Deane said thoughtfully.

The policeman turned on him. 'That's enough questions, sport. It's *my* job.' The last good humour went from his face and it was set.

Deane stared at him, then stepped aside. He shrugged his shoulders, ridding himself gladly of the affair. He

20

looked at the thin body, sprawled under the dirty canvas in the back of the truck. But now it was only dead meat; it was done.

The policeman's eyes travelled across the street. 'You'd better come back inside and make a statement.'

They walked back into the tin box of the station again. Deane's feet dragged reluctantly. The sight of the pen and the report sheet gave him an unwelcome stirring.

He thought of a charge-room in Sydney, derided himself once more for a bloody, cockeyed fool to start this going . . .

'What's your name?'

'Deane. John Deane.' He supplied the real one.

'From?' The policeman was short-spoken.

'Down south.'

He raised his square head. 'That's no crow-eater's accent.'

'I'm English.'

'Pommy, eh?' The policeman looked him over, stained hat and worn trousers, unshaven face and dark, sprouting hair and sweat. 'What's your business in the Territory, pommy?'

'Passing through.'

The grey eyes observed him. The policeman scratched his chin. 'You'd better tell me about those jokers who picked up the abo.'

Deane described the men briefly, the big white man, the aborigine stockman, the other two. Then he finished; that was the end of it.

Silence waited. The policeman twiddled his pen between his fingers. The room was heavy and airless. At last he said, 'All right. That's it then, Deane. Leave it to me.'

Deane asked, 'Recognize the men?'

Slowly, the policeman shook his head.

Deane stayed silent. He went outside.

The policeman followed, and they stood under the overhanging, hot tin roof. He turned to Deane. 'I'll get the body off the truck for you. Then you'll be done with it.'

'All right.'

'Anything you need before you pull out?'

Deane stared at him. 'Pull out?'

'Yes.'

Suddenly he knew that he wanted to go, and the policeman wanted him to go. His eyes narrowed.

He said, 'Where to?'

The policeman waved his arm ahead down the trail.

'Where's the next halt?'

'Anderson's station. Thirty miles. Bonzer homestead.'

Deane jerked a thumb at the Austin. He was bitterly amused. 'In that? I've got both front springs gone.'

'There's no garage here. You'll get no repairs.'

'I can't shift until I do.'

The policeman chewed his lip. He was too anxious to see the back of Deane: there was something smelly in the air.

Finally he said, 'Strap it up with mulga shafts. They'll hold.'

'Not me—I've tried that. And repair shops come fewer still ahead, I guess.' Deane was keen enough to get out of it: but no sense in getting bushed. A man

could die, stranded in that kind of country of the Northern Territory. He said obstinately, 'I'm staying here
until I get it fixed. I can have a couple of replacements
radioed up. You get a mail plane here?'

'Once a week.' The policeman spoke shortly. 'At
the airstrip at Clancy Rock.' He turned to Deane again.
'But the next trip isn't till Friday following.'

Deane stood in the strong heat of the sun, looking
out across the street, turning the thing over. He wanted
to move on; but it was better to stop.

He said, 'I'll wait.'

'I'd advise you to push ahead.'

Deane summoned up the saliva to spit fiercely and
defiantly into the dust. 'I'll wait.'

'It's up to you, sport.' The policeman chewed his lip
again. Then he swung away, stepped down to unload
the small body from the truck. Then he carried it round
to the back of the station, wrapped in a cloth.

Deane got into the driving-seat, eased the Austin
forward very gingerly. He parked it at the back of a
cyclone-ravaged shack, among fallen timbers and packed
earth. Then he brushed himself down, wiped the sticky
sweat from his forehead.

He swung his bag over his shoulder, strode loosely
over the hard ground towards the hotel. He saw the
policeman leave the same building, return across the
baked street to his office. Deane turned behind, and
now the policeman was back standing at his own door,
watching.

The loungers on the veranda, the woman outside her
store, the children and the solitary aborigine: these too

23

watched Deane pass. The afternoon was late; long black shadows came behind him as he walked, sliding over the ground. The hatbrim shielded his face but the sun was still harsh and cruel.

They all watched him, and he looked straight ahead as he went.

Strident rays of light slanted over the hotel roof. The bar was closed and the door was locked. Deane went round to the back, where the crude privies stood in the unfenced yard. He pushed through the rear entrance of the hotel, went down the passage.

The hall was bare and unfurnished, whitewashed, with wooden benches along the walls and a door marked 'Bar'. Corridors led off to the other apartments and the dining-room. Deane stood still. No one came: then he called out. 'Hullo there.'

A man stepped through from the veranda and walked slowly across the hallway. He wore singlet and faded fawn trousers belted tight at a lean waist; he was thin and scrawny and he had a lopsided, battered face. He stared closely at Deane. 'Yes?'

Deane said, 'You're the manager?'

'Yes.'

'I want to stay here a few days.'

The man shook his head. 'Sorry. Can't do anything for you, mate. Just happens I'm full right up.' He looked Deane over, gazed past him down the passage.

Deane began to laugh. 'Here? In a spot like this? What've you got on—a race meeting or something?'

'Full up.' The man made a gesture. 'That's all.'

'Listen.' Deane stepped forward. 'I'm not asking for

a suite and bath. Just give me a couple of yards on the veranda.'

The hotel man said, 'I can't take you, mister. Sorry, but nothing doing. Try somewhere else.'

'Where?'

'That's up to you.'

Deane came over to him. He was tired and fed-up. 'All right. Joke over—I'm staying. Now show me where I can unload my bag.'

'Nowhere.'

'Show me.'

The man looked at him silently. Then he said, 'Have it your way . . . Sign in the book.'

Deane followed him down the corridor. The man flung wide a door: the bare room opened on the veranda. Several low beds were spaced out.

He said, 'Pick your own bunk.' Then he went out.

Deane called after him. 'I need a wash.'

'Wash?' The man half-turned in the dark corridor. His unfriendly face became almost amused as he laughed. 'The only wash you'll get is in beer. Don'cher know there's a drought on?' Then he went forward out of sight.

Deane settled on a bed in the corner, flung himself down. He was burnt dry, dirty and sticky and drained of strength by the sun. He wiped himself over with a towel, lay back for a while in the shade. Sunlight fell outside. The street stayed silent. Dust streamed in the air, filtering down. There was no sound of movement in the hotel. He could feel the heat radiating from the iron walls.

He lay still and wondered what was wrong, around this town.

After a few minutes he shifted, dragged his legs forward and got up. He wandered down the corridor to a wash-house; it was empty and without water, faintly sour-reeking. Deane turned aside, went back to the airless room. Then he realized he hadn't eaten for a long time. He crossed over the hall to the dining-room.

It had walls of whitewashed corrugated iron like the rest of the building, wooden tables. Two men in check shirts sat at a table; nobody else was in the room.

Deane dropped down at one of the other tables. The men looked at him speculatively. They had finished their meal: after a while they got up and walked out. Deane sat there alone.

The door to the kitchen opened at last and a girl's head peered round. She came into the room. 'Good evening, sir.'

''Evening,' Deane said.

The girl was young and pale-skinned. Her face was anxious. She asked hesitantly, 'You want?'

'Tucker. Something to eat and drink.'

The anxious expression deepened. 'I am afraid it is late—' She had a faint foreign accent.

The door behind her was flung open. A woman stood there, red-faced, glowering. 'Tea's finished. We close at seven.' Then she slammed the door again.

Deane examined his dusty watch. Ten minutes to. He looked up at the girl. He said slowly, 'I haven't eaten all day.'

She paused. Then she said, 'I'll see . . .'

She went back into the kitchen. The door closed. Deane heard voices raised, loud and strident, argument, then calm.

After a while the foreign girl came back. She put before him a plate of cooked beef. Deane relaxed slightly. He said, 'Thank you.'

The corners of her mouth lifted in a half-smile. She had a grave face. Then she fetched a helping of steamed jam pudding, a cup of strong tea.

Deane ate quickly, swallowing the hot food. Outside, the temperature fell lower as the sun went down, and the iron walls began to lose their stifling heat.

He was alone; voices came from behind the closed door. He felt very much on his own, a stranger in a strange town.

When he had finished he went out, walked exploratively around the small town as twilight fell. In the morning he intended to have the spare springs ordered by transceiver radio. When they came he'd get out surely enough, fast as he could.

Nobody was about now. The town was dead. Husks of empty buildings, bat-infested shells. Smokes of fires rose straight above the aborigines' camp outside the town. The dry, sickle-shaped leaves of the gums lay still in the breathless air. In the creek the earth was caked hard as rock; and beyond was the scrub, the mulga bush and then the hills. Tension began to pass slowly from Deane: the hard white light of the open street was gone, the watching eyes had disappeared. Sunset flushed pink, then swiftly red as red dust, darkening, and the evening was solitary.

Lights came on in one or two of the inhabited houses, the post office, the store. The café, where a crackling radio played out to the night; wooden cubicles, but nobody sat inside. Flickering oil-lamps, kerosene lamps. Then as the night fell, the land softened despite its dead rainlessness and sterile soil.

Scent came on the night, the leaves began to stir. In his hardness Deane was conscious of that softening. Some of the heat and the weariness eased from him. For a short while he forgot the fugitive flight that had brought him here, the harshness he carried with him; and the incidents of his arrival. He felt safer under the night sky.

He forgot the aborigine and the hanging, and his own abrupt pity for terror under a hot sun. Then as he walked back through the dark to the hotel, there was a rush of wheels through the dust behind him. A vehicle swept by, braked hard, pulled up outside the police station.

It was a Land-Rover. A burly man got out, crossed over the wooden entrance and walked into the office where the policeman was still working. The lamplight caught him clearly: this was the same man whom Deane had met that afternoon in the bush; the man who had taken away the aborigine to murder him.

And Deane knew suddenly he would find himself caught up in that affair after all.

3

HE stood cautiously on the edge of the dark. The night was colder and he shivered, almost trembling with a mixture of violent anger and despair.

He looked in through the bright doorway of the police station. The policeman had risen to his feet: he and the newcomer were talking. The policeman was showing no suspicion or hostility towards the big man; rather he was deferential, waiting on his words, nodding. The man in the tailored grey jacket had his back to the door and his shoulders were set wide and square.

Deane retreated into the shadows. Slowly he walked on to the hotel. He said again harshly, Nothing to do with you, mate. Not your blasted concern. Just get out . . .

But he knew he couldn't go. And then suddenly something else resisted in him, more dangerous although it made the rest. Justice: although he didn't know what sin the abo was supposed to have committed, although it was a useless word anyway. Justice: for a terrified black who never seemed to have received any formality of justice.

He knew it as a half-conceived impulse, springing mainly out of resentment because he too was an underdog, on the side of the undertrodden, the abused.

Savagely Deane shook his head, and then the caution came back to him.

He was still arguing with himself as he pushed his way into the hotel, sprawled on his bunk. Starlight was cool through the window shutters. He lay still under the net, glad to forget it; until sunlight brought morning.

He went out into the new day, crossed over to the police station. Now he was restless inside, not sure which way he would jump.

He walked over the deserted street. He remembered how he had walked with the same tight grimness through the deserted morning streets of Sydney's Kings Cross on his way out, waiting for the cars to come sliding at the kerb to pick him up: the cars which never found him. They'd have given him no justice, no more than the poor damned abo got . . .

At the police station he paused; then he said, 'Sod them all.' He went in.

The room was empty. There was a step behind, and the policeman followed him inside.

He looked Deane over. For all his grey hair he was stalwart in the khaki uniform. He wasn't pleased to see Deane; a frown flickered, was smoothed off.

Deane said, ''Morning.'

'G'day.' The policeman jerked his head. 'What is it?'

'I want to order up those spare springs.' He spoke abruptly.

'Got the truck details?'

'Here.' Deane pushed over a scrap of paper.

The policeman took it. 'All right. I'll have it sent through.' He had a square, ordinary face but his expression was shuttered. He turned away, refilling his pipe with strong fingers: the bowl was fitted with mesh guard for easy lighting in the bush. Behind him the walls were spread with maps, posters and typewritten notices.

Deane asked, 'You've got a transceiver here?'

'No.' He spoke reluctantly. 'It's packed up—broken tubes. But we're using the one at Clancy Rock. I'll get a message over to them today.'

Deane said, 'I'm in a hurry for the stuff. I want to catch the morning transmission. I'll take it over to the station myself if that's the quickest way.'

'You've no truck.'

'I'll get a ride. Or borrow a vehicle.'

'No need for that.' The policeman's voice was sharper. 'I told you I'd handle it.'

'All right.' Deane turned aside. He glanced back. 'Any developments?'

'Developments?'

'Concerning the aborigine?'

'None.'

Deane said suddenly, 'What was his name?'

'Thomas Clancy.'

'Shouldn't be so difficult to find out who killed him, in a lonely kind of district like this.' But he had little doubt the policeman knew already.

The copper said, 'I told you before, that's *my* business, pommy.' His voice was thicker.

Deane shrugged his shoulders and walked out.

31

He stood in the sun, gazing over the township once more. The faded black paint on the hotel sign, *Galah Creek Hotel*. The café, the couple of houses; the store, *Thompsons*. And the dry, sandy earth which filled the fallen shacks, the broken windows; the township which was dying to the bush, like others of the Territory he had passed through. The days were gone of the cattle-droving, the mining and prospecting; and the abandoned towns were falling to the sand . . .

Hot tin and weatherboard in the sun, burn of light, still air and sharp, clear vision to the horizon, over the town and beyond to the trail. A few trees and a few shacks, a dry creek set in an open plain. A scattering of white men and an aborigine camp. An unmade road in and an unmade road out, past the long miles of the cattle station paddocks; and then the silent bush all round where the last nomad tribes lived, the emptiness.

A mail plane coming in once a week, and the trans-ceiver radio as the only other link with white men and women beyond. A tight community, hanging on to life against the death of their town: but death in the wilderness would come.

Let it come: no concern of his.

He began to walk down the street. The children who were playing in the road stopped from their scuffling, stood still and watched him pass. The old aborigine who sat under a corner-post of a veranda lifted wrinkled, screwed-up face to him. The vast silence of the township, its heat and aridity, oppressed Deane. There were no teenagers, no young people: a few, ragged children, mostly half-caste and black: lean, dried cattle-hands: and the old. A town without future, cracked wooden flood-boards creaking under his feet. Only the hot sun was live, pressing down.

Impatiently he pushed open the door of the post office. It was a wooden shanty like the rest; inside there was a counter, a few G.P.O. posters and lists of regulations fixed on the wall. The room was dusty and old reference books were piled on a shelf. A set of tarnished brass scales lay at the far end of the bench. All was silent as the falling dust.

A man came in from the rear section of the building. He stood behind the counter, examining Deane. He was fleshless like a rake, desiccated, a loose singlet hanging over bony shoulders, loose slacks; and he was nearly bald. He said flatly, ''Mornin'.' When he spoke his mouth revealed he had only a couple of teeth left, top and bottom.

Deane asked, 'When does the mail plane come in again?' Even if there hadn't been any reason for a lie, he didn't trust the copper.

'Friday.' The man's eyes held just the same cautious suspicion Deane had seen on the policeman. A small town: but no outback hospitality for the stranger. They had something on their minds.

'Where's the airstrip?'

'Clancy Rock.' The mail-man's voice too was curt and careful.

Deane said, 'I want to go out there today. How can I make it?'

The man behind the counter paused. Then he shook his head. 'Walk it, I reckon.'

33

'I thought perhaps I could get a lift.'

The thin man shrugged. 'That's up to you. Lucky to find any traffic goin' that way.'

'Anybody got a car I could borrow? My truck is broken.'

A tight smile touched the bony face. 'We don't many of us run cars around here. Ain't nowhere to flamin' well go.'

Deane's own smile flickered in return. He glanced around. 'Good job you've got here. Not much work, heh?'

'I got enough.' The man stared across. 'I handle all the stuff for the station, y'know. Freightage, the rest. Keeps me busy enough ...'

Deane nodded, turned away and came out.

A certain interest began to hold him, awareness. He knew something, and he was ready to pry to a bit farther. He wasn't just thinking of the abo now: he wanted to give himself a kick.

He had run a long way nonstop in fear and bitterness. Now he had to halt a few days. And he was ready to bite back a bit, make some trouble here or anywhere else.

Not a nice man: a foul-mouthed bloody shiralee perched on his own back.

And perhaps he would make somebody sorrow about a brief incident of abo-killing, after all.

The next building held the café. Deane slung open the door, walked in. He stood by the counter; behind him were the dark wooden cubicles, empty. The counter carried a plate of beef sandwiches and a tea urn. A small man with a moustache got up from a seat.

34

Deane said, 'Tea—please.'

The small man poured out a cup. 'You take it with milk?' He was a Greek.

Deane sipped the tea, standing by the counter; it was strong, and the milk was tinned. The radio played music softly, crackling with bursts of static. He said, 'Know anyone who might be going over to Clancy Rock Station today?'

The café proprietor had brown rich eyes; but suddenly they were shuttered and remote as the eyes of the policeman, the hotel-keeper, the post-office man . . . He shook his head. 'Nobody much goes over there. Why should they? Sometimes the boys come into town. But not ver' often.' He moved his shoulders, regretfully.

Deane said, 'The copper's going over. But I don't fancy a lift from him.' He grinned.

The café-owner smiled back, very slightly.

Deane asked, 'What's his name?'

'The constable?'

'Yes.'

'Lawrence. Mister Lawrence.'

'Been here long?'

'Many years. Many years . . .'

'His beat covers quite a bit of the country around, I suppose?'

'Vast miles. But this is his home.'

Deane nodded.

The Greek stood stiffly behind the counter. He fiddled with a knob of the radio, then glanced back. 'Passing through, heh?' He was still cautious.

'Soon as ever I can.' Deane said, 'What's this fellow

Malone like—out at Clancy Rock? Big man, square-shouldered type?'

The small, moustached Greek turned the radio knob again. Static howled across the shack. Then he said, 'Mister Malone's a big man, yes. Big man, indeed, an' in lots of ways. It's a fine station he has, many thousand head of cattle.' He became respectful. 'One of the biggest men in the Territory, and very rich.'

'I'm beginning to guess that,' Deane said. He put down his cup, came away.

A blaze of light seized him in the street. His lonely feet paced hollowly down the sidewalks. At the entrance to the store he turned off, went inside.

His eyes adjusted again. He ran a crumpled handkerchief over his face and wiped away the sweat, standing in the shade. He glanced around him, over the walls of the room: piles of tinned foods, ironmongery, cards of buttons, spades and dishes. Bolts of cloth, barrels and socks.

A woman came through from the back. He had seen her the day before, standing outside on the front steps when he had ridden in. She was middle-aged, stout as one of her own flour-barrels in contrast to the dried, skinny men of the district. She wore a yellow print flowered dress which was shapeless like her own potato sacks; and it was tied with a man's leather belt at the waist. She wore battered pink bedroom slippers and she had fat legs. She looked him up and down. 'What can we do for you?'

But her voice was cheerful and her eyes were frankly open; replacing the doubt and suspicion he had met before in this town.

36

His own mood changed to meet her.

He smiled softly. 'You've got a lot of stock around here, missus.'

'That I have.'

'Don't tell me you ever sell it, in a forsaken dump like this.'

''Course I do, you cow.' She was loud in good-humoured indignation. 'Think I'd perishin' well keep it otherwise?'

Deane jerked his thumb at a hanging row of print dresses like the one she was wearing. 'You're not going to convince me there's that number of women around Galah. Haven't seen any of them if so—they must be hiding from me.'

'Don't get excited, mate.' She laughed back lustily at him. 'Not unless your taste runs to dark meat . . . They're for the station—Clancy Rock. They supply 'em to their lubras, y'know. And Edwin Lawrence draws 'em from me too when he hands out to the tribal women on Ration Day. Every litt'l helps . . .' She slapped her fat flanks. 'Got 'em all sizes, you can see. Never found a lubra yet I can't fit—from the kids to the fat old mammies. They don't ask no Paris cut, you can bet . . .'

'What about the white women?' He watched her, grinning.

'Which white women?' She winked at him. 'None around here, matey. Only me, who don't mind abo-dress because she gets it next to free. And the Czech waitress down at the hotel, and she don't wear much either. That leaves Grace Fisher whose old man runs the

37

hotel, and Malone's wife at the station. And they never touch my stuff, I'll tell you. They mail-order fancy things up by plane, from way down south. Don't ask me why, when there's nobody to look at 'em . . .'

Deane was taking it all in. He said, 'I suppose Malone's station means a lot to you here in Galah. Brings business, all the rest?'

She looked at him shrewdly behind her raucous *bonhomie*. 'You bet it does, matey.' Suddenly she was direct and keen-eyed. 'Without Clancy Rock we could just about pack up. No boozers in the pub, no mail for the office, an empty café. No bastard kids to shout in the street. No orders for me . . .' She heaved her fat shoulders. 'No flamin' water even, 'cause we bring it all in from the wells on his land when the soak goes dry. Without Clancy Rock we're a dead duck at Galah, mister . . .'

She stood with her plump arms folded, watching him. She was more intelligent than she looked in her shapeless dress and bulging heavy legs. She watched him now as carefully as the men had done, and she too knew what load he had brought into the town . . .

Suddenly Deane was flushed with warm exultation, a flood of blood in which it was all mixed: awareness of the added risk he was taking on, spouting of vengeance, heady power: the worm in the rotten apple, the stranger in the street, man with a load of mischief. It was like spice, like wine, the knowledge that he held. To use it: worth all the danger. To be someone important, no fleeing down-and-outer here: to make them fear him.

He stepped back, satisfied and elated now that the picture was coming clearer. Then he said casually, 'I dropped in for a new belt. I used my last one to lash up my truck springs—it did it no sweet good.' He jerked his head at her waist. 'Got one?'

She lifted her hands and moved back, laughing again. 'Not this one. Hey, not this. Me knickies'd soon be down.' She gave him a sly glance. 'And don't say that'd please you, you dirty young feller . . .' She went behind a cupboard and fetched out a brass-buckled stockman's belt. 'Suit you? Take this one?'

'Dinkum.' Deane slipped it round his sagging trousers.

The fat woman watched him. 'Ah, if I had a waist like that. Not for thirty years. Too much bloody beer, pommy. Too much beer and not enough exercise . . .' She winked.

Deane said, 'How do you know I'm a pommy?'

'God—a stranger in a town like this.' She laughed again, shaking her wobbly cheeks. 'Think *anybody* doesn't know? . . . Anyway, you've got it in your voice at times. Quite the la-di-dah, eh?'

Deane said softly, 'I'm no gentleman, dearie . . .'

'No?' She was still laughing.

He paid. Then he said, 'I want to go to Clancy Rock Station this morning. How can I get there?'

She stood with a flabby hand on the rail of the dress-rack. Sunlight through the window caught her round, pink face with its bulges of fat. She paused, her face settling. Then she said, 'I'm runnin' a few things over for their stores. Maybe I'll take you.'

39

'When?'

'Now, mister.'

'Thanks,' Deane said. He tightened his belt another notch, followed her out to the step. He looked back. 'Aren't you going to lock up?'

She threw a glance at him. 'We're a funny kind of bunch in Galah. But we ain't thieves.'

'Maybe your visitors are.'

'We keep an eye on 'em.'

Deane smiled. She lowered her fat bulk into an old Holden with an open truck body. He sat beside. 'Thanks, Mrs.—Thompson, is it?'

'That's my name. Margie Thompson.'

The car bounced over the ruts, started out along the trail by which Deane had come in. He glanced over the town as they moved by. Sunlight poured down. The place was empty as ever. A couple of dingy dogs, the children. Empty windows in the shacks.

They passed the shattered bulk of the courthouse. He said, 'Some building.'

She drove fast and heavily, never dodging the bumps, bobbing up and down in the seat and plunging steadily on. 'That's one of the old ones. Not many left, what with the willie-willies . . . This used ter be quite a town, way back. Had a copper mine a few miles outside. And the droving-track went right by. There was some life then, lusty men and their laughin' sheilas . . . Those days are gone, pommy . . .'

He glanced at her profile with the sagging chins. 'Been here all your life?'

'No bloody fear.' She spoke loudly above the engine

noise. 'I came all the way across from Perth with me old man, over twenty years back. We was restless. Don't ask why we stuck here, don't ask me why. 'Cause the end was due to come, even then. But we did . . . Then he got crook, kicked the bucket . . . I stayed on. My place now, an' it suits me. But I ain't got my roots so deep as some of 'em . . . I can pull up and shift if the time comes. I'm not worried . . .' She looked back at him, with the same shrewd acuteness.

They drove on, and soon Deane watched the fencing wire come back into sight again, the dry station paddocks. He said, 'Pretty bad drought, heh?'

'Pretty bad? Too right it's bad. It's a fair, stinkin' cow. Worst drought I ever knew . . . Many more, and this district'll end up in the sand . . .' She whistled cheerfully as she drove, an old bowl-shaped hat pulled down over her head against the sunlight. Then she turned to him. 'Open up for us.'

Deane climbed out and opened the gate in the wire, closed it behind. They went on, down the fork towards the homestead.

The fat woman said, 'An' what about you? What's your story, pommy?'

Deane didn't answer. She turned to look at him again. At last he said, 'Nothing much. I like the roving life . . .'

'Where are you going?'

'Nowhere special.'

'You'll find nowhere easily enough, round about here.'

'Maybe I'll reach Darwin, ship myself out.'

'On this road? Off the bitumen? Take you months—even if you don't do a perish on the way.'

'I'm in no hurry,' Deane said.

The sun was higher in the sky and the heat was stronger. The paddock was bare and leafless, dry scrub and scorched soil. The nakedness and aridity under the sun began to affect Deane, burning through him with the heat and sterility. The woman felt it too, falling silent.

The car rode on. Then they saw cattle ahead, spread out over the paddock in small groups.

He asked, 'Malone losing many head?'

'I'll say he is.' Her voice was quieter. 'Just look at 'em when we get nearer. You'll see all you need.'

'No water?'

'Ain't the water, pommy.' She too stared over the paddock. 'Always some in the well. But when the drought's on the feed gets eaten out. Poor flamin' cattle have to go further and further away for it. They get weaker and weaker. In the end they're not strong enough to get back to the bore. That's finish.'

They drove on. The atmosphere began to oppress Deane more heavily. Nothing in sight, bare earth and the groups of cattle standing, motionless. The harsh sun poured down over the plain, a white ball of light. When they were close he saw the beasts were mere skeletons, filthy hides draped over their scarecrow backs. They stood gaunt, legs splayed out, unmoving: spectre-like symbols of doom and destruction. Only their horns were smooth and unchanged; their great eyes had a glazed, vacant stare. They did not see, they did not hear the car pass. They just stood, trance-like.

42

The woman glanced at Deane. She stopped the car, switched off the engine. Then the silence was something frightening. Absolute, utter silence over the vast paddock and desolate earth. No wind, no movement, no life. The dying animals with the empty eyes. Time stood still, waiting for death.

Deane shifted uneasily. One of the cattle nearby went down with slow finality into that long-promised death. It sank to its knees, rolled over with the same complete silence to the bare earth. A nightmare, death-watch quality hung over the scene.

The woman stared at the sky, looking for the kites and eagles which would arrive. Then she started up the car, drove on. They moved forward through the herd of dying animals.

Deane swallowed dryly again. The woman sat tightly in her seat. When she spoke her voice was stripped. 'Drought in these parts does something to you, pommy. When it's on the land, it changes things.'

'I'll guess that.'

A dead bullock lay across the road. She steered round. Her voice stayed expressionless, the bawdy life gone from it: 'Death's about you, and maybe you get more than a litt'l mad. And more than a litt'l dangerous, fightin' to live for yourself.'

'Don't we all?' Deane said.

She lifted a hand from the wheel and pointed forward, through the stricken land. Roofs vibrated in the killing light. 'Well, that's it, pommy. That's Clancy Rock Station. Richard Malone's place.'

Deane peered forward.

43

She repeated, 'Remember it. People gets a litt'l mad, a litt'l dangerous.'

'That sums me up too,' Deane said.

4

THE Holden pulled up to the front of the homestead, stopped. Deane swung himself out and stretched his legs.

He stared around the station. 'Fine place.'

The fat woman nodded.

The big homestead was single-storeyed, with fresh white walls. Around the veranda were yellow hurricane shutters, and pots of green plants were growing in tubs. Farther away the outbuildings were heaped, sheds and huts; the low whine of a diesel generator came from the distance. The homestead was newly-painted, gloss enamel still shining in the sun. Gardens surrounded it, couch grass lawn . . . The drought laid its hard hand but it could not kill, here . . . A sprinkler fountained water through a pipe, over the lawns. Spray drifted against Deane's dusty face.

He looked back, and away in the distance were the death-plains, the falling cattle.

The woman climbed out of the car behind him, heaving and sighing with the exertion. 'Wait till you see inside.'

They crossed over to the stores sheds. Deane helped her carry the boxes they had brought; her heavy hips swung in front of him, pinched by the leather belt.

He kept his eyes open in case he should see the man he knew was Malone. But the only people in sight were a couple of aborigines: an old, white-haired man in thin trousers, an unhurried gardener: and a young black house-girl who crossed over from an outhouse to the homestead, carrying a basket.

It was quiet and peaceful, even a graceful scene; only the high, bright sun was hard. Water drifted down on the grass, catching rainbows in the sunlight.

Deane felt a taut wave of impatience breaking over him, fierceness. He put down the parcels at the entrance to the store. 'I'm going back to the homestead. I want to get fixed up about that radio message.' He turned away from the fat woman.

She didn't reply. He walked off, striding across the soil. He stared at the house: but there was no sign of the big man of the Land-Rover ... He paused on the steps, then went up to the veranda.

Inside, he stopped still, staring round again. A living-room: a heavy-piled carpet of rich blue. Long gilt mirror, mahogany furniture. Scarlet and white curtains to the windows, a cabinet with china and silver piled high within ... An air-conditioner beyond the mossie-nets.

Through the door, a refrigerator, deep-freeze cabinet side by side. An automatic washing-machine in the kitchen ...

Beyond the expensive rooms, the watered lawn, beyond in the full heat of the sun lay the skeletal cattle; and the stifling bush and the dust and aridity, and the naked, ancient aborigines. The pale white walls made a box, enclosing a rich unreal scene ...

Deane smiled grimly. He stepped farther down the veranda. A man came through one of the doors, faced him. He gazed curiously at Deane. He was thin and tall, his face highly-strung.

Deane said easily, 'Who are you?'

The man hesitated. Then he said, 'I might ask you that.' He had a mild voice, pitched quietly. 'If you want to know, I'm Welles. Mr. Malone's manager. Where are you from?'

'Galah Creek.' Deane said, 'I want to send a message over your transceiver. I've got a broken truck—have to order up some new springs. Who handles it?'

Welles held out a hand. 'I'll take it. I'll be getting through soon.'

Deane passed over the slip of paper he had written out. The other man studied it; then he looked up. 'All right... I'll tell them to put it on the next plane. That's Friday.'

'Thanks.'

'Travelling through, are you?'

'That's it,' Deane said.

'How did you get over here?'

'Got a lift from Margie Thompson at the store.'

The station-manager nodded. Then he said, 'Fix yourself a drink or something before you go back. We can't help you with a truck or spares, but if there's anything else we can do just say the word.'

Deane stared at him; then he asked, 'Mister Malone around?'

'He's out in the paddock.'

'Doing what?'

Welles' expression altered again, becoming surprised,

46

annoyed. 'Shooting some of the cattle. We're losing them fast . . .' His voice was sharper. Then he said, 'Why?'

Deane heard the crack of a distant rifle as he listened. 'Just thought I'd like to meet him.'

Welles' face was puzzled. He opened his mouth to speak. Then a woman stepped through from the kitchen to the veranda, and he turned to her.

She said indifferently, 'Hullo, Philip.' Her voice was tired. Then her eyes came round to Deane, opened fractionally.

Deane said, "Morning.'

Welles explained, 'He's stuck in Galah with a broken truck. I've promised to order up some spares for him.'

'Have you?' she said, nodding. She stood in the doorway, still and disinterested, even bored. But Deane's eyes moved over her with a kind of greed, because this was a lonely land without women . . . She wore a clean white dress and she was still youngish, early thirties maybe, and she had a fine full figure which swelled out the white cloth which wrapped her.

He swallowed and pulled his eyes away. Then his gaze came back to her face and she was good-looking, fair and fresh-skinned, with just that mood of disinterest, or maybe only resignation, in the half-closed grey eyes.

Welles paused, uncertain. His thin face looked more anxious. Then he said, 'This is Mrs. Malone. This is Mr.—er—"

Deane said, 'Deane's my monniker.'

'Good day, Mr. Deane.' She shaped the words, unemotionally.

47

'Hullo,' he said. He added, mock-politely, 'I like your homestead, Mrs. Malone.'

'Do you?' She nodded, remotely, as though she were thinking of other things. She had a deep voice which showed little of the Territory accent, interesting in its very passiveness. Deane's eyes flickered over her body again and then he wished he were a mile away, beyond the itch of temptation.

'Makes a change from Galah Creek.' His voice was thick, talking roughly.

'I suppose so,' she said.

Welles commented, 'Galah isn't much, these days. Fading out, like most of this area.' He glanced over the yard, at the outbuildings.

The soft pattering of the water continued on the lawn. The intermittent, dull sound of rifle-firing came from far across the paddock. The sun drenched the land with bright, stifling heat and light.

The woman said abruptly, 'It's too hot to stand here. Let's go inside.'

Welles spoke quickly. 'I promised to send this order off for Deane. I'll take it down to the radio now.' He walked with jerky strides along the steps, over to the far end of the building.

'He's in a rush.' Deane stared as he disappeared out of sight, then his eyes came back to the woman.

'Yes.' There was the slightest smile. 'Anyway—come in while you wait.'

'All right.' He watched her body sway as she moved forward, and he followed her into the carpeted room with its mahogany furniture. The nets dropped like

48

silence behind; inside, only the air-conditioner purred softly.

Deane stood in the room, dirty and rough and stinking with sweat. He stood stiff.

The woman said calmly, 'What's the matter?'

'Don't think I'm bashful. But my condition isn't exactly suited to this place.' He still spoke thickly, fiercely, because he was angry with her for the lush setting, for the snare of her own sex.

'It doesn't matter.' She said, 'I'm not houseproud. We can always buy a new outfit if you defile it . . .' Another small, wry smile touched her lips.

Deane sat down sprawling in an embroidered chair. He ran a hand over his bristly face and looked at her. She leaned against the cabinet of china and glass. The white dress tightened over the line of her stretched leg and disturbance ran through him again. His tongue moved over his lips, wetting them, and again he stared down at his broken, dusty boots. He thought of the touch of bare flesh, and his dry throat closed up.

She said, 'Philip Welles certainly went away very hurriedly. I think he's a little frightened of me.'

'Why?'

'He's parted from his wife. I think he—' Then she stopped. His desire grew, and at the same time he called her a bitch.

He stared at her face again. Arched fair eyebrows and corn-coloured hair, a cool, isolate face. A wide mouth with pursed lips which could maybe contradict that isolation.

Then she looked back at him and their eyes met full for the first time. Something changed in her. She moved

49

abruptly. She stirred again, then seemed to uncoil, releasing herself.

She said, 'Want anything to drink—or smoke?'

'Not now.'

She straightened herself to her feet, walked slowly to the window, looked out over the station, the paddock. Shen she came back, leaned against the cabinet again. She shifted restlessly, then was still.

Deane said across the silence, 'That's some tasteful china you have there.'

'We bought it from London.'

'You're a collector.'

She said, 'Not really.'

Somehow their voices altered, became strained. His eyes roamed, came back to her. Once again their glances met. His lips began to split in a hard, desperate smile.

Slowly, and almost helplessly, she began to smile back.

She leaned awkwardly against the cabinet. She was rigid now in herself, beneath a pale, burning face. She gazed away, then her eyes came back. Finally she wrenched her gaze from him.

Deane watched her incredulously, excitement suddenly leaping. Swiftly he felt hot blood surge in him, expectancy beginning to rise and pound.

He spoke with the same thick tension, his voice distorted. 'Lived here long?'

'Five years.' She opened her eyes to him again, showing rueful hopelessness. Suddenly she was open, giving intimately.

He swallowed. 'Had too much of it?'

'Too right.'

'The bush?'

She nodded, shaking her head uselessly. 'Emptiness. Then drought and death...' She pointed outside with a bare, white arm.

Deane never took his eyes off her, watching with a hungry intentness, knowing she was conscious of his stare. Now she was breathing more quickly, her face flushed. He had forgotten all about Malone and the rest.

Sense of power seized him again, the new ecstasy of the conqueror. She looked at him freely. There was a noise outside, along the corridor, sound of feet.

Her eyes despaired, allying themselves with him. She said, 'My lubras—in the kitchen.' Again their eyes searched one another. Suddenly she said, 'It's too hot here.'

He nodded, motionlessly waiting.

She said, 'Let's go outside.'

Without speaking, he followed her. They went down from the veranda, across to one of the outbuildings. He strode behind her, his mind empty of any other thought. Nobody else was in sight.

She went into a small outhouse. It was thatched with dried spinifex grass. Inside, the walls were covered with the grass. Piped water ran down them, evaporating, cooling. The cooled air struck against Deane. The room held a couch and a chair.

The light was dim, pale green inside. She stood in the room, turned to look directly at him. The white dress clung about her. Her face was still flushed and once more her eyes met his.

Deane stepped forward blindly, almost choking.

51

Breathlessly they seized one another, tumbling on the couch.

When he came round he moved his head, stared at her. After a while he raised himself. She got to her feet.

They dressed themselves hurriedly, without speaking. Deane buckled his belt, turned back to look at her again. Her face was held away in the shaded light as she zipped her frock.

Abruptly he jeered. 'Love at first sight, heh?'

She did not reply.

Then he softened. He stood silently in the room for a while, watching her dispassionately, a kind of understanding coming slowly. She smoothed the couch level again, straightening up, brushing aside her tangled hair. At last he said, 'We'd better go back.'

She said, 'You first.'

He went out, returning to the veranda. The homestead was still deserted. After a minute she came out of the shed and walked back towards him. He watched her come under the hot sun, tall and well-made, crumpled white dress, strange, good-looking face which wore a mask that had slipped. She lifted her eyes as she came, regarding him silently; she stepped up to the veranda beside him, beginning to pass by.

From the other side of the buildings a car hooted stridently. She stopped, and Deane stared across, and the fat woman from the store was back in her seat in the Holden, waving. He waved back, shouting. 'Coming.'

He stared at Malone's wife. Then he said, 'I don't want to walk home. And there goes my transport.'

She didn't speak.

He started to move, and then he stood still. He said, 'Ever come into Galah?'

She stayed silent. Then she said, 'Sometimes.'

'Tomorrow afternoon?'

'Perhaps.' She spoke flatly.

'Somewhere quiet.' He said, 'See you round the back of the old courthouse. Then maybe we'll find out each other's name . . .' He smiled briefly, came away.

He crossed over and got into the old car beside the fat woman.

She started up the engine. 'Wondered where you'd got to. Everything go off all right?'

'All settled, Mrs. Thompson,' he said.

She looked at him speculatively, her gaze passing over his rumpled clothes and untidy hair, his unbuttoned shirt. Casually he smoothed his hair and fastened the buttons, eyeing her in return. Then the car jerked forward, heaved itself on the return track.

Suddenly he remembered what he had come to this place for. His mood was complex now, unsettled and confused. He stared through the windows of the car, across the paddock. When things got mixed he wanted to run out, retreat and escape as he'd always fled when 'complications came . . .

Another vehicle was travelling over the paddock towards them, rousing its dust-track. He remembered how he'd seen the truck approach across the bush, the day before. And this too was a Land-Rover. Suspense began to tighten up in Deane's stomach. The vehicle came to meet them, travelling to the homestead on the same track.

A black wearing a loud check jacket sat in the front seat: the black stockman he had met on the plain. And the driver was the hard-voiced white man. The fat woman braked the Holden, pulling over, and the cars came together, moving slowly.

She waved a hand, beaming. ''Morning, Mr. Malone.'

The big white man nodded back. His gaze moved over, encountered Deane. Recognition came into his eyes.

Deane sat tight. He was still muzzy-headed from fornicating with Malone's wife: now Malone had something dangerous on him too. For a moment he was still, showing nothing.

Then a flash of the old desperation took him. He leaned across the truck and raised his hand. He saluted, grinning. 'Hullo, Malone. Strung up any more blacks this morning?'

Malone's dark brows bunched, lowering. Then the Holden lurched forward abruptly, into the dust-cloud, back on the route to Galah. Behind, the Land-Rover started moving as well, in the opposite direction.

The fat woman pulled her hat down over her head and drove faster. The easy smile had gone from her face and it was anxious. Then she said, 'You shouldn't say things like that, pommy. Won't do anybody any good. No good at all . . .'

5

DEANE shrugged. 'Maybe it won't. We'll see.'

'You're a fool.' She steered the wheel, and now her expression was harsh and unfriendly like all the others in Galah. 'A pie-eyed fool . . .'

Deane stirred in the seat, twisted with brief anger. 'Why the hell? I know he yanked that abo up high—I'll bet you know it, and everyone else knows it. Why shouldn't I show it to his face? Murdering bastard.' He spoke rapidly.

She said, 'Isn't as easy as all that, pommy.'

'Why not?'

She was silent. They reached the scattered herds of cattle in the paddock. Silence, again the waiting, the glazed stillness before death. The beasts had not moved since the car had driven out: fixed, caught, oblivious during the time of his lust, death waiting near squirming life . . . Deane brushed his shirt-sleeve over his sticky, sweating face. He stirred again, disquieted. He stared away from the plain, looked back at the fat woman.

He repeated, 'Go on—tell me then. Why should Malone get away with it? There's every sign he will, with that spineless policeman.'

'Edwin Lawrence isn't spineless.' She pressed her plump lips tighter together.

Deane banged his fist sharply against the hot metal panel of the truck door. 'No?'

'No.'

'Then why doesn't he do his job?'

'Because he doesn't want to see Galah die, any more than the rest of us do.' She spoke hurriedly.

Deane said, 'I've no special love for the abos. I've never had anything to do with them. But I guess they're men, like anyone else.'

'The fine, upright young pommy, searchin' for justice, eh?' She laughed. The untidy hatbrim came down over her eyes. The wispy hair straggled out beneath. Her fat, loose cheeks shook with the car's rattling motion. She stared ahead through the windscreen.

'That's me,' Deane said.

'Not so upright,' she said. 'Been takin' the opportunity for a sniff around the boss's wife. Haven't you now?' She didn't turn her head.

Deane's voice came savagely. 'Mind your own bloody business.'

Suddenly she took her foot off the accelerator. For a moment he thought she was going to brake and chuck him out of the truck. But it was only a brief reaction; then she sighed. The vehicle drove on, more slowly.

At last she said, 'You ain't been around here very long. You don't know how it is.' Her voice was tired; she too wiped her face.

Deane asked, 'Well—how is it then?'

She said, 'Listen, pommy. I don't know what Richard Malone did or what he didn't do. But I told you before, this kind of country makes a man act in funny ways. And a drought like this finishes it all off . . . He's had a lot of trouble from some of his blacks lately. They start chuckin' their weight about. There comes a time one of 'em gets uppity, something goes . . . You teach him good and proper.'

'It's still murder.'

'Sounds good. Sounds easy.' She was contemptuous. Then fresh seriousness touched her. She glanced across, taking her eyes off the uneven track. 'Listen, matey. You've seen a bit of this corner of the Territory. You know there's plenty more blacks than whites—more station and township blacks even, without considerin' their muckers outside in the bush. Once you let them get too big for their flamin' precious boots—that's it.'

Deane said scornfully, 'Think they ever would? Not the specimens I've ever seen. Not a poor scared bastard like Thomas Clancy . . .'

'Just try it.' She glanced back at the road ahead, in time to steer sharply around a vast rut. The chassis crashed down, up, surged onwards.

She added, 'Anyway. You know how it is. If anything happened to Richard Malone—or if he took a dislike even to us poor jokers in Galah—you know what'd come, pommy.' Her voice had gone flat.

'Well?'

She paused. 'We need him—but he don't need us.' She watched him levelly. 'He don't need to use the hotel to provide drinks for his boys. He don't need the café, he don't need any piddlin' litt'l store—why, he keeps a bigger one himself. He could handle the mail himself, the banking—hell, he just sticks to Galah 'cause it suits him, 'cause he likes to know there's a litt'l town near by. Galah's always been here, an' maybe he's sentimental. But we're like flamin' pigeons feedin' out of his hand. Flick, and we're done, pommy.'

Deane looked at her. She watched his hard eyes and her face appealed to him defiantly. He shrugged it off again. 'We'll just have to see,' he said.

The truck pulled down into Galah Creek, stopped outside the fat woman's store. Deane eased himself out, stood in the open street, drew on his belt.

He grinned down on her. 'Thanks for the lift, Margie.' She sat, still watching him; and he turned away, strode down the street.

He met the policeman coming across the square. Deane stopped him. 'Don't bother to send that message about my springs. I've been across to Clancy Rock myself.'

The policeman tautened warily. 'When?'

'Just got back.'

The policeman paused. Then he shrugged. 'Righto.'

Deane said, 'I saw Mister Malone over there.'

The policeman did not speak.

Deane said, 'I guessed it. He's the fellow from the Land-Rover. He's your man, Constable Lawrence.' He smiled softly.

The policeman paused. Then he said, 'Thanks for the information. I'll have a talk with him.' He turned away, stuck his pipe back in his mouth and crossed over to his office.

Deane stood alone in the square, still smiling.

58

Then he felt hungry, and he walked across to the café for something to eat. He fed in solitude, sat for a while, and afterwards he went back to the hotel in search of a drink, cold beer.

He stepped in through the front door. The white-washed passage was lined with benches; a couple of men squatted there, drinking fast. Two others stood in the narrow bar, glasses tilted. They looked like cattle-hands, check-shirted and dungareed.

The hotel-keeper stood behind the stained wooden counter. A few bottles were placed on a shelf behind him; there was a kerosene refrigerator with chipped and faded white paint. The bar was dark and drab, and the iron walls radiated waves of thick heat.

Deane went up to the counter. The two men were talking to the hotel-keeper; at the sight of him they stopped.

Deane said, 'Beer, please.'

The hotel-keeper hesitated. His shiny face was uncertain. He glanced at the two cattle-hands.

Deane waited, growing impatient. 'Beer . . .'

At last the man took a can, jerked the cap off. He passed it across. The bar was stifling hot, low-ceilinged, a dark hole. Deane swallowed the iced drink. Then he lifted his head from the glass, glanced around the bar. The hotel-owner was leaning back against his shelf. His thin singlet was damp on his narrow chest; his fingers twirled a can-opener between them.

The two cattle-hands were lean, dry fellows, hard-bodied. One of them was younger than the other; he wore green shirt and pushed-back hat, and he stood with

59

his feet apart, regarding Deane. He said, 'Goin' to shout us a grog, pommy?'

Deane stared back at him. 'It's a pleasure.' He put his hand in his pocket and pulled out a few coins.

The hotel-keeper opened the cans silently, poured out. The cattle-men lifted their glasses, drank.

Deane asked, 'You from Clancy Rock?'

The cattle-hand in the green shirt said, 'Yes.'

'Good station?'

'Dinkum,' he said. 'Fair dinkum.' He added, 'And we're goin' to keep it that way.'

'Why not?'

'Heard you was over there this mornin'.'

'That's right.'

The other cattle-man drained his beer, put down his glass. 'My turn.'

Deane lifted his own half-full tumbler. 'Save it till next time, thanks.'

The young man in the green shirt puckered up his mouth in a grin. 'The pommy don't like too much beer. 'Fraid it might upset his constitution, old boy.'

The older cattle-hand laughed with him.

Deane grinned back tolerantly. He finished the beer, began to walk out. 'So long.'

'Hey!'

He turned. 'Yes.'

'Want to talk to you.' The man in the green shirt stepped forward.

'Well?' Deane stood still.

'You made a crack at Malone this mornin'.'

'What do you know about that?'

The cattle-hand rocked on his heels. He had taken a lot to drink and it was affecting his eyes. The other man had come up beside him. The younger one said, 'You'd better move out pretty stinkin' quick. We don't wancher in Galah.'

Deane snapped back. 'I'll get out as soon as I bloody well can—don't you worry about that. I've seen all I want to see of this perishing place.'

He walked through the doorway, from the veranda to the ground. He took a couple of paces, then swung round.

The two men from Clancy Rock were coming for him. The younger one thrust forward. Deane ducked and kicked his ankle roughly as he went by: the cattle-hand dropped on the ground.

Then Deane stepped aside again as the other man reached him. Again he kicked out. The two cattle-hands pulled themselves up, grimacing: they came forward more carefully. Deane moved swiftly, broke through, swung about again. The younger man caught at him: Deane fought savagely because he was desperate. They mixed it for a while and then the hand tumbled back. He fell on the ground, clutching.

The other man clashed with Deane. Now Deane had lost his temper. He struck out hard, used his boots again as the cattle-hand slipped back. The brown-faced man fell heavily. Deane staggered, held himself upright. Blood spurted from a tear on his cheek. The two men from Clancy Rock sprawled on the bare earth. Deane put his hands on his knees, bent down and panted.

At last he straightened his back, lifted his gaze. The hotel-keeper watched from the veranda. The other men had gathered near him. Then a loud, throaty voice said, 'That's enough, Deane.'

Deane turned, wiping away the blood. The policeman Lawrence stood behind, angry-faced.

Deane said, 'Enough? I should think it damned well is. They jumped on me—'

'Leave it, Deane. I'm blaming you—this is the first blue we've had here for months. I don't stand for fighting on my doorstep. You're a trouble-maker.' He had the khaki police hat pulled down, and hot malice showed in the eyes beneath.

Deane shook his bare head exasperatedly. 'Blast you —you saw.'

'Don't swear at me.' The policeman drew himself up. 'You'd do best to get out of Galah. There's no place here for you.'

Deane felt the fury surge rapidly. Then he swallowed it back. He muttered, 'Sod you.' He pushed past, went up the steps into the bar. The hotel-keeper followed him in. Deane jerked his head at the bottles.

He swallowed the beer. Then he went through to his room, flung himself down on the bunk and slept heavily through the afternoon heat.

At last he got up, yawned, ran his fingers across his scalp which was still stubbly like a reaped cornfield from the stockman's crude haircut back at Kooni. He walked down the corridor to get another drink: beer, water, anything wet: he had sweated a lot, and now he was thirsty as a horse again.

He got a cupful of water. Near the kitchen he met a woman he hadn't seen before. His eyes ran over her:

she was too heavily-fleshed, a big ripe woman with a thick waist despite the slick city-style dress she wore: she was made-up a lot. But it was a sensually-attractive face, strong and bold, and she looked at him freely.

Deane said, 'Hullo.'

'Good afternoon.' Her voice affected something like a sophisticated accent. She bore a swirl of perfume around her. 'You're Mr. Deane?'

Deane said, 'I am.'

'I'm Grace Fisher. My husband and I run this hotel.' They stood in the corridor. 'Sorry to hear you've had some trouble. I hope you'll like your stay here, Mr. Deane.'

Deane's smile twisted at the phrase. Her big, slightly cowlike eyes looked at him in the dim light. She opened the door of her room and the sunlight splashed out behind her.

Perfume drifted over him again. She leaned backward, her body arched. 'You're from the Old Country?'

He nodded.

She said, 'Well anyway—let me know if there's anything at all I can do for you, Mr. Deane. Anything at all.' Her eyes were bold again despite the polite gentility, almost hungry, and another pulse of sudden expectancy stirred in Deane.

'I'll let you know, Grace,' he said. He went on down the corridor, feeling extra pleased with himself again. Not so many ripe plums in Galah: but those that were seemed ready enough to fall to the hand of the stranger ...

He went out, on the street. As he passed the old aborigine who lounged against the corner-post, the man lifted his crinkled face and spoke to him. 'Boss. Aw, boss . . .'

Deane stood still, staring down. The old boy wore ancient black trousers and a battered black coat without a shirt. He had frizzy hair and a remarkably ugly face. Deane said, 'What?'

The old abo stared down the street, then up to Deane. 'Hey, boss.' His voice dropped confidingly. 'You bin friend of blackfeller? You bin friend, eh?' He watched Deane.

Deane stared back. Then he said cautiously, 'Friend . . . Maybe I'm a friend, maybe not. Why?' He bent over the aborigine.

The abo sat forward, drawing nearer. He screwed up his face under the filthy old hat, conspiratorially; but he didn't smell so good. 'You bring 'im in, dat feller body poor ole Thomas? Dat feller policeman you no like 'im. You bin havem plenty hard words, eh?'

'Me no like 'im—not much.' Deane grinned briefly.

'You friend of blackfeller—eh?'

'Just you say what you want.'

The old aborigine hesitated. 'You no tell 'im, boss. You keep plenty quiet, eh? You no tell 'im other mans?'

Deane said, 'Tell what?'

'You no tell 'im, what this feller speakem you. You keepem blackfeller's secret, boss?'

Deane moved impatiently. 'I shan't talk. If you've anything to say, say it. You tell. You tell, plenty quick . . .'

The old aborigine paused again. Then he said, 'Awright, boss . . . Listen, boss, listen . . . Dat feller Malone, he try to kill 'em, them two blackfellers in bush. Dat feller Thomas he hung 'im up, that udder fellow Jack he got 'im away. Dat feller Jack, he run 'im away plenty quick when big boss 'e hang 'im up first feller . . . Boss he shoot 'im, but Jack he bin get away. He bad hurt—he hide 'im out longa bush.' His deep eyes stared up.

Deane looked at him, sorting his way through the gibberish, thinking back: and he remembered the second scared aborigine who had sat in the back of the Land-Rover that afternoon in the bush, covered by the rifle of the other white man. An abo named Jack . . . He said, 'This feller Jack—what's his other name?'

'Jack—Jack Clancy. Him too from Clancy Rock.'

Suddenly he realized. 'This feller—he saw Malone string up Thomas? He saw him kill Thomas Clancy?'

'Yes, boss. Him saw . . . But 'im real crook, out in bush. Him wantem help, boss. Him bad . . .'

Deane paused, turning it over. An eye-witness, even if it was an abo. Maybe an ace up his sleeve. Bitterness moved him again. He'd pull the place down about him if he had to . . . Then his gaze hardened. 'Why you tell me?'

The aborigine looked down the empty street again, up at Deane. 'Blackfeller need help. Can't bring 'im pore blackfeller into town—boss Malone bin lookin' for 'im, could be he find. Can't tak 'im back longa station. Sick feller him bad, not last longa bush.'

'What about your friends out there? Blackfellers in the bush—why can't they look after him?'

The old abo shook his head. 'Dem bush fellers not bin 'ere. All gone walkabout—maybe dey come back, bimeby. Dat feller Jack all lone, boss.'

'Where do I come in?'

'Him way out longa bush. Needs truck, boss. You not belonga this town, boss—reckon you help pore blackfeller. Could be . . .'

'Why should I?'

'You not like these white fellers round'bout here. You good man, boss, eh? Help pore crook blackfeller, not let Boss Malone 'e catch 'im.'

'You flatter me.'

'Okay, boss, okay?' The abo raised his eyes again.

Deane thought it over. He was still suspicious. Then he said, 'Blackfellers settle their own affairs. They not call 'em in white man.'

'Yes, boss, yes.' The old abo nodded fiercely, rolling his eyes. 'Blackfellers dey needem help, truck. Needem white man, stand up that bad feller Malone. Blackfeller no good, Malone 'e catch 'im plenty quick. Blackfeller longa this place afraid. You helpem, eh, boss?'

Deane still watched him warily. He took another glance over the deserted street. But he had nothing to lose. At last he said, 'All right. I'll go to see this feller Jack Clancy. You showem way, heh?'

'Awright, boss.' The abo nodded vigorously again, smiling. Another wave of stale smell from him passed over Deane, and he held his nostrils.

He said, 'What time we go?'

'You bringem truck, boss?'

'I'll find one.'

'Longa sundown, boss. You come over blackfellers' camp? No one see 'im?' His skinny arm pointed across the creek.

'All right. I'll come across.' He said, 'This feller Jack —you say he got shot. Where?' He gestured to his body.

'Him shot longa belly. 'Im bin bleedem plenty.'

Deane stood up. 'All right. See you later.'

He walked on down the street. He wasn't sure exactly where it was all going. But if he could get that ace up his sleeve, trump card to play . . . Malone must be wondering pretty hard where the second abo was hiding . . .

Suddenly he walked back to the old aborigine, bent over him. 'Just you tell me,' he said. 'These two black-fellers—Thomas and Jack. What'd they been doing? Why boss feller Malone him try to killem?'

The abo stretched slowly, spat in the dust. He leaned back against the ant-eaten post. 'Boss feller Malone, 'im bin goin' after blackfellers' womans. Him bin leavem own wife, catch 'im blackfeller girl. Dis feller Thomas, him no like 'im, him tell 'im . . .'

'No account for tastes. I'll stick to white women . . .' Deane said, 'All right. Incidentally, what's *your* name?'

The old abo said, 'Me Albert Decla.'

Deane got up, walked forward again to the store. He pushed inside. The fat woman was unpacking a box. He said, ''Evening, Margie.'

She turned round, surprised. Then she hesitated, smiled back. 'Hullo, Johnny.'

'Going to do me a favour?'

'Any time,' she said, looking coy.

'Not you too,' he said. 'Not tonight, Josephine . . . Will you lend me your truck for this evening?'

She observed him shrewdly, hands on her big hips, the sack-like dress falling about her. 'Where do you want to go?'

'Just a-riding.'

She said at last, 'All right. Behave yourself.'

'Much obliged, Margie. I'll do that.'

'In more ways than one, pommy.'

'I'll do nothing I shouldn't.' He walked back to the hotel. On the way he went round to the rear of the shack where he had parked his own Austin truck, collected the first-aid kit he carried.

Then he strode into the hotel for his evening meal.

He sat alone in the heat of the stifling dining-room. The foreign waitress brought the same old meal. Cooked meat and potatoes. Steamed jam pudding, strong tea . . .

He sat back afterwards, loosened his belt. It was some time yet to sundown. He lounged idly, watching the waitress clear the table.

She was a slight, fragile little thing in a black dress, with high cheekbones and violet, luminous eyes. Dark-haired, with white skin, incongruous in this rough and derelict outback hotel. Her pale white hands moved to and fro, lifting the plates, the bent cutlery, and he watched them curiously.

At last he said, 'What are you doing here?'

She turned to glance gravely at him. 'I work here, sir.'

'I know that. But what brought you to a dump like this?'

She explained seriously, 'The job was offered. So I came.'

'How long ago?'

'Four years, sir.'

Deane said, 'Hell—forget the "sir".' Nobody but a foreigner had ever called him that, in Australia. He asked, 'You're a New Australian?' He smiled with a touch of sarcasm.

'Yes.' The small white hands rested on the edge of the check tablecloth.

He said soberly, 'You should have stuck to a city. This is no place for you.'

She threw out her chin. 'It's all right. I'm not fussy— I am content.'

He liked her soft, faint accent. 'Okay. It's up to you.' He shifted his legs. 'You're a Czech?'

'Yes.'

'Why did you choose Australia?'

She lifted her shoulders and dropped the corners of her mouth, shrugging as though it were a foolish question.

'What do they call you?'

'Elsa.'

'And the rest?'

'The rest?'

'Of your name.'

'Elsa Lehman.'

He nodded. She picked up the plates, swung away and went to the kitchen. At the door she glanced back. She smiled again, with a shy, uncertain charm. Deane lifted a hand and waved back. Then he stubbed out his cigarette and got up.

He walked out into the street, paced slowly down to the store, and picked up the Holden. He checked the petrol and water, started up.

He drove through the town street to its end, across the dry bed of the creek, across to the other side where the aborigines' camp was erected. He drove carefully down to the pile of heaped huts and wurlie dwellings, beyond the blue gums of the creek. The sun had set low and the air was pink, darkening.

As he went he wondered what he was letting himself in for, this time.

6

THE dwellings were built out of old bits of boarding, of iron, broken doors and scrap. Piles of rusted tins and bottles lay on the trampled earth. Fires smouldered in crude hearths, guarded by strips of corrugated iron. Mangy cats and a couple of scabby dogs roamed beside the huts; a handful of aborigines stood near their sheds.

The sun was down. Deane picked his way through the camp. Some of the wurlies had bark roofs; on top of one of these was an abandoned meal, a chunk of dirty kangaroo meat, smothered in flies. A small black girl nibbled a bone. The eucalyptus gums of the creek marked off the area from the white man's town. Beyond lay the open bush, quiet and still.

The old aborigine Albert came to meet Deane. He held himself upright, his thin, scrawny legs tottering in small steps. ''Lo, boss,' he said.

Deane said, 'Let's go,' his lips pressed together. They

went back to the truck, and the old man climbed in stiffly beside him. He carried a can of water and a great chunk of half-raw meat, and he smelled worse than ever.

The truck moved forward, bumping over the rough ground. 'Longa dis way,' the abo said. 'I show.'

They drove over the dry creek, circling the town. Deane said, 'Where is he, Albert?'

'Longa Clancy Rock.'

'The station?' Deane turned sharply.

The abo shook his frizzy head. 'Naw, boss. Longa rock—Clancy Rock.'

'Where's that?'

The abo pointed with his bony arm, the black coat tightening, felt bursting from the torn seams. 'Dis way. Little bit not far from station. But dat feller Malone, 'e no find 'im.'

Deane said nothing. He stared ahead as the Holden rattled along over the uneven ground, running parallel with the main track.

After a while he asked, 'You-fellers, you plenty angry Malone, eh?'

The old abo grunted. Then he said bitterly, 'We town-feller boys. We soft now—not gottim no fight. White man he treatem rough, killem, we say nothin'. But dose bush-fellers, when dey come back, dey no likem. Dey got plenty fren's along town-boys.'

Deane shrugged. He steered through the paddock gate, on to Malone's land. The long, empty plains lay ahead.

The old abo turned to him again. 'Me peaceful feller, boss. Wantem see dat feller Jack all right, no trouble.

When dose bush-fellers come back bimeby, could be plenty trouble. Could be, boss . . .'

Deane glanced at him, a little scornfully. The old abo had pulled down his shapeless felt hat against the wind. His lined, stubbly face beneath the brim of the hat was puckered and solemn. Deane said, 'Malone—he ever acted like this before?'

The frizzy grey head shook vigorously. 'Naw, boss . . . Dat feller, him gone proper mad. Could be him gone bush mad, maybe . . . We see, bimeby . . .'

'We surely will,' Deane said. Dark was coming down. He glanced behind, but the bush was empty. The land was dead still under the drought, hushed in night. Groups of silent bullocks, glazed eyes glinting red in the light of the truck's headlamps. Once, a quick gleam in the dark, green and flickering, vanishing; dingo. But the rest of the plain was dead, lying out barren under the great stars.

They came to a fork in the track. The abo said, 'Longa there, boss.' The Holden drove on. Air streamed past Deane's head. Something rose upon the clear horizon: closer, and it was a steep rock outcrop, standing out from the bush. The dark bulk broke the flat surface of the paddock, shadowy, black and white with starlight on burnished stone.

Deane asked, 'Clancy Rock?'

'Yes, boss.'

'Where's the station, Albert?'

The abo gestured to the east.

'How far?'

'Little bit not far, boss.'

72

Deane stared around. 'He's up there?' The mass of rock looked hard and comfortless an abode, deserted.

'Yes, boss.'

'Supposing that feller Malone, he sees our tracks?'

'We tak 'im away, boss. We hide 'im, that feller Jack, plenty quick.'

Deane drew back. 'There's no other place. He'll have to stay here for a while.'

The old abo stared at him. Then he pointed the way. The truck jolted over a broad stony patch, desolate as a gibber flat. They came to the base of the outcrop, parked in its shelter.

Deane switched off. The rock loomed over them, hiding the sky. Night was silent. The dark lay heavy. Then he got out. 'Where is he?'

'I showem, boss.' The abo led the way up the rock face, stumbling, clutching at the can and joint of raw meat he had brought. Deane followed with a torch and first-aid kit. They climbed slowly, picking their path over the sloping falls. To one side was a crevice, a dark slit. The abo shuffled laboriously towards it. At the entrance Deane looked back, and the bush was spread out a hundred feet below, black and still to the horizon.

The crevice formed a small cave. Deane swung the torch, and an aborigine lay inside, stretched out on the rock. A grey blanket covered him. His eyes rolled in the light; his face was sweaty.

The old abo went forward and spoke. Deane bent forward to examine the wound. He said, 'You Jack Clancy?'

'Yes, boss.' The injured abo's voice was a whisper. In the thin yellow beam he looked sick.

73

'You saw that feller Malone, big boss, him hangem up Thomas Clancy? You saw, him kill 'im?'

The man stared. His face cleared and he nodded. 'Yes, boss.' Then he shivered, far down the length of his body. The cavern was draughty and bare.

'All right, Jack. You good boy.' Deane opened the first-aid kit.

The aborigine had bled a lot. Deane put a dressing over the holes where the bullet had gone in and come out. No chance of a doctor for a long way round, and the transceiver was the only contact. They could get in touch with the F.D.S.—but only from Clancy Rock Station . . .

Deane stood up, hesitating. A faint wind which moved over the rock blew down the crevice and was cold on his scalp. The old abo looked at him, waiting. The wounded man lay quiet; Deane recognized him now as the man who had sat in the truck that hot, burning afternoon, guarded by a rifle . . . Now all was cold and dark, and the black skin shone wet, and the fierce, primitive face was haggard.

He thought everything over. If they carried the abo back into Galah Creek, it would get round to Malone's station and the business would come to a head, one way or another. He wasn't ready for that. He stared at the man again: the aborigines were a tough crop and could take a lot. The abo would be able to hang on for a while on his own, till the moment came.

He said to the old man, 'This boy, him not too bad. Him be all right, longa couple days. We come back, eh, Albert?'

The old abo said doubtfully, 'Him real crook, boss.'
'He's okay.'

Albert still looked uncertain. Then he shrugged, bent and put down the hunk of meat and the water-can beside the wounded aborigine.

The abo lifted dark eyes to Deane for a moment. Suddenly his were like the despairing eyes of the other aborigine back on the plain: on the edge of pleading and held back only by pride. Then the emotion died as the other had died before; the man's eyes dropped, the eyelids closed, and his head turned away.

Albert spoke to him once again. He pushed the stale meat nearer. Then he came away and Deane followed, down the rock slope to the Holden.

Deane drove away slowly, over the stone flats which which would hide their tyre marks, back to the road.

The truck picked up speed, and they drove towards Galah.

Deane stared behind, at the harsh and lonely rock. He shivered, contemplating what it would be like to remain alone there all night, sick and weak. He glanced in the direction of Malone's station, and realized the rock would be the likeliest place for anyone to search, were they looking for a wounded man across the bush ... He stirred guiltily. He masked his discomfort by fierce driving, swinging the wheel.

Albert watched him, glancing sidelong. They came back within sight of Galah, skirted it. Deane braked harshly. The old aborigine lifted himself out.

Deane said, 'Meet you again tomorrow. We arrange something later, heh?'

Albert shifted his feet in the dust. Then he said, 'Awright, boss.' He turned silently, and walked away over the dry soil, towards the camp. His figure disappeared into the dark. Deane drove the Holden back to the store, parked it and returned to the hotel.

He was finishing his breakfast the next morning when there was the clump of boots down the passage, and the policeman came in. He stepped over to Deane's table.

Deane raised a casual hand to the khaki figure. ''Morning.'

The policeman nodded. He watched Deane with the same careful suspicion on his brown and creased face. 'Come over to the station when you've finished, will you? Want to talk to you.'

'All right,' Deane said. Lawrence turned away and walked back down the corridor.

Deane drank his tea, went outside into the sunlight; he crossed over the street once more and entered the tin shack of the police office.

The constable stood inside, and upon a straight-back wooden chair sat Malone. He looked up as Deane came in, and they stared at one another.

Lawrence said, 'Richard Malone and I want to have a talk with you, Deane.' He was stiff and upright, and his eyes wandered with disfavour over Deane's unkempt, untidy figure and ill-shaven face beneath the cropped stubble of hair.

Deane's gaze went back to Malone. Malone got to his feet, and he and the policeman waited beside one another. Deane saw again how strong the station-owner was, his shoulders and heavy body, the good-quality jacket and trousers, expensive touches of gold ring and neck ornament; and the assured, arrogant face of the big man, the boss, the lord within his estate . . . And there was cold dislike as he watched Deane.

Deane said, 'You and Mr. Malone?'

'Yes.' Lawrence stepped forward. 'I spoke to Malone here when he came into town this morning. And he agrees with your story that you met him when he caught up with Thomas Clancy. But he's explained that Thomas jumped out of the truck again and ran off shortly afterwards. He never saw him again.'

'Is that so?'

'That's so.' Malone spoke calmly.

Deane said, 'Glad to hear it.'

The policeman was thumbing the bowl of his pipe. 'I wanted you to know, Deane. Just to clear up any misunderstandings.'

'So maybe old Thomas strung himself up?'

'All right, Deane.'

Deane was thick was anger. He began to turn away. He stared at the copper, and his lips twisted in a sneer of contempt.

Something flickered in the constable's face: renewed animosity, and then a kind of shame. He dropped his eyes. Deane swung round, and went out into the sunlight of the square.

A voice came from behind him. 'Deane.'

He looked over his shoulder. 'Well?'

Malone walked down the steps. 'Come and sink a beer with me.'

77

'I don't want one.'

'Come on.' He strode across, hat brim pulled down, big and burly. He spoke as though he offered forgiveness, a great favour.

'The bar's shut anyway.'

'They'll open for me.' The swell of power came into his hard voice.

Deane lifted his shoulders indifferently. They went round the back of the hotel into the dark bar. The manager came quickly and poured out the drinks.

Malone said easily, 'Thanks, Jim.' He pushed across the beer to Deane. 'Here.'

Deane took a swig, put down the glass. Again he and Malone looked at one another.

He felt no particular emotion except curiosity. He studied Malone dispassionately, eyeing him from three different angles and seeing three different men: the big boss man who owned and ran this district: the murderer who had hung up the aborigine: the husband whose wife Deane had known.

And Malone looked back at him with cold glinting eyes.

Deane said, 'What's this for?'

'What?'

'The drink.'

Malone said, 'They've been talking to me about you, pommy. I thought it time we had a few words ourselves.'

'Who's been talking?'

'Some of my friends about the town.'

'You've got no friends here.'

He watched Malone fire immediately, then push the temper down. 'You'll see.' He had a harsh voice, which was without polish but not uneducated, and strong. His jaw was square and solid, his eyebrows heavy. He looked a hard, driving man with a devil of violence in him.

Deane saw he had red rims to his eyes, gleaming in the shadows like those of his own crazed bullocks . . . He said rudely, 'You always call the tune, Malone. That doesn't make for friendship.'

'I call it, too bloody right.' He swirled his glass, the liquid frothing; then he swallowed the beer with one long draw. He put down the empty glass, licking the froth off his lips. He watched Deane with added dislike. At last he said, 'So you're passing through, pommy. Stuck with a crook truck. And you're moving on when the plane drops in the spares, Friday.'

'That's the position.'

Malone said slowly, 'Better keep to it.'

'Why?'

He leaned forward suddenly from the bar counter. 'Listen, you perishing son of a whore. You bastard abo-lover . . .' He spoke in a voice which started low, rose.

'Go on.'

Malone said, 'You blow off a lot of bull about me. That I don't give a damn about—you don't matter any-how. But when you start on other things—'

'What sort of things?'

Malone fastened hot eyes on him. The diamond pin in his neckerchief winked in the light behind the bar; his elegant attire was incongruous against the hard face with blue-skinned chin, the bloodshot eyes. 'You're a cursed

79

stranger. You don't know this country. But I own it. I own it.' His voice was louder. 'A quarter the size of your bloody Old Country, and remember that. This is my land, and when I find any abo-lovers on it—'

'Well?'

'I get them out quick. Pretty damn' quick.'

'I'd be only too pleased,' Deane said: 'I'm no abo-lover, Malone. I couldn't honestly say I care a tinker's cuss about them. You've got the wrong idea . . .'

'You think that?'

'Yes.'

Malone shook his big head. 'You're out to make trouble. You're a lousy, quarrelsome joker with a grudge, a useless clown who's made a muck-up himself and envies everyone above him. And that includes the whole flamin' town for you, pommy. I've found out all I need to know about you . . .' He thrust himself nearer. The hotel-keeper hung in the background, half out of sight, listening carefully.

Deane said, 'I'm no beauty. I'm an unpleasant sod—let's agree to that. But it wasn't me who killed the abo . . .'

Malone jeered. 'It wasn't. So . . . ?'

'I think maybe I let the poor blighter down.' For a moment he was sober.

'You? Malone laughed. 'Who's going to put any trust in you? Think he couldn't see you for what you are? Think anyone couldn't? You look a useless bloody cow, a crapping down-and-outer—and you flamin' are one.' He dropped his heavy fist on the counter abruptly, clenching it. 'And you—you walk in with your load of trouble, you try to make mischief—you try to throw your blasted weight about—'

Deane pushed the drink away. 'Keep it.' He began to get fierce himself, coiling up.

Malone's eyes were flaming red. Then he pulled himself back. 'Listen.' He banged the table again. 'Till that cursed plane comes and you get to blazes out of it —just be quiet. Just be quiet, pommy. Take it easily. I'm telling you.'

'You're telling me.' Deane stared at him, at the flickering, unstable lights in his eyes. He wasn't sure whether to get amused or violent in his turn. He chose to get amused. And then he was bored. He turned away, strode to the door.

Malone called out. 'Just you hold it a moment.'

Deane swung round. 'I've had all I want.'

Malone was leaning forward from the counter, staring. 'Something else. What were you doing in Margie Thompson's truck last night with that dirty old abo? Where did you go?'

Deane said, 'We went for a ride in the dark.'

Malone's eyes glinted again. 'Don't talk to me like that.' His voice came slower. 'I don't care what the hell you say about me, pommy. There's nobody to hear you. So just yap, like a crapping little cur . . . But don't go mucking about with the abos. I'm telling you.'

'You said that before.'

'You better listen.'

'I don't feel inclined.'

Malone came forward, stopped a couple of feet away. The heavy body was planted on two sturdy legs, set

81

apart. He swayed, stood still. Then he said, 'I'll warn you once more. This is my place and I'm keeping it. I'm not letting it go, either to the blacks or to the bush. And most of all not to the blacks, the stinking barsteds...' His voice was quiet now, but in it was something obsessive, more dangerous than his loudness. He said, 'Everything's headin' for finish. The whites are leaving this district, headin' for the towns like flies to a stinking heap. Galah's dying the way my bloody cattle are, falling. And the abos are waiting to come in—station blacks, bush-fellers, they're all the same. Just a handful of us, and the bunch of them breedin' out in the bush, spawnin' more of them in their camps. But I'll keep 'em down, I'll hold 'em back. It won't happen in my time. I'll stop them coming in . . .'

Deane watched him intently.

Malone said, 'And if anyone encourages them—God help him. God help him . . .' He was only whispering now, his gaze fixed on Deane.

Deane shrugged his shoulders. He jabbed a bit farther. 'If that's how you feel about them, you'd better keep away from their women in future, Malone. Otherwise you'll find your own bastards pushing you out.'

Malone grunted. He dropped the wide shoulders and his head came down as though he were going to charge. The small eyes sparked. Then the fury was smoothed instantly from his features. He straightened up again, altering. From red temper his voice became bare. 'I know which side you're on. I know how to treat you from now on.'

Then he swung aside. He kicked away a bench to

82

clear his path, viciously, and crossed over to the bar. He turned his back on Deane, ignoring him.

'Jim!' he called out sharply.

The hotel-keeper bobbed his head in. 'What can I do?'

'Beer,' Malone said. 'Have a grog with me, Jim.'

'Thanks,' the bar-keeper said. They swallowed the drinks together, knocking them back fast.

Deane stood at the back of the room; Malone took no further notice of him. At last he turned and went outside, leaning against the veranda-post.

He waited in the sun, gazing across the street and considering. His gaze fell absently on the half-caste kids and the few passers-by, on old Albert against his corner-post and the lean dogs in the shadows. A Land-Rover came round the corner of the shacks and pulled up near the hotel.

The black stockman was driving. He glanced up at Deane; then he propped his feet on the bonnet and sprawled back, waiting.

Later Malone came out of the bar. He walked past Deane without speaking. He strode over to the Land-Rover. The stockman jerked himself alert quickly. He started up: Malone got in and the truck moved away, down the main street towards Clancy Rock.

Deane scratched his chin. Now he knew Malone better. A man with an obsession, unbalanced: and a man like that you couldn't knock down and settle it; he'd keep coming back at you, again and again. You had to kill him and make finish. Or else take to your own heels and run.

Then he walked down the street, past the black

shadows of the buildings, towards the store. He went inside.

The fat woman was cleaning out her kitchen at the back: a small, old-fashioned place with an ancient black range. She straightened up, polishing cloth in hand. 'Hullo, pommy.'

''Morning.' Deane leaned against the wall, regarding her.

She said, 'Want anything?'

He shook his head. 'No.' He hesitated. 'Just feeling rough.'

'Why tell me?' Then she stared at him closely. 'You sure don't look so perky right now. Haven't even got that sour go-to-hell look on your dial. Not goin' crook, are you?' She put down the rag and stood in her untidy frock, fat arms hanging down at her side.

Deane said, 'I'm fed up. Thoroughly damn' fed up with this spot of earth. It gets me down.'

'What's so wrong?' The sun from the window behind her outlined her body in the loose dress, the great calves.

'I've just been talking to Malone.'

Her stare sharpened. 'He's in town? Well?'

Deane said, 'He's a bit bloody up the pole, isn't he? Round the bend somewhere?'

The fat woman looked at him. At last she said, 'Ain't we all? I told you that before.'

'How long's he been this way?'

'We all know we're in a dyin' town, a perishin' district . . .' She spoke softly.

'He's flaming dangerous. He might do anything.'

'Why tell me?' she said again, wearily.

'He runs the whole place.'

At last she said, 'Malone's all right. Nine-tenths of the time. He's got a down on the abos, and sometimes it throws him off-beat a bit. The way I told you.' She mopped the shining sweat from her brow.

Deane muttered, turning aside. Then he looked back. 'So why does he chase after their women if he hates them that way?'

'I guess men don't always take a jump at women because they love them. Eh, pommy?'

Deane moved restlessly. He went to look out of the dusty window, over the yard and privy, the crosses of the sand-covered graveyard beyond.

He turned back to her, considering her with her wispy brown hair and fat face, loose body and floral dress, shuffling-slippered feet. And the clearness of her blue-green eyes in the swollen middle-aged body, staring back at him.

He said, 'I'm a rotten type, Margie. I only want to revenge and destroy. Never to build. You know that?'

'Yes.'

'I want to attack this town. Attack Malone—pull him down. I fool myself it's for justice because he killed a poor bloody abo and nobody seems to care about that. But it isn't the reason. Not really.'

'Then why is it?'

'Probably just because I've got a grudge. I don't know . . .'

She kept looking at him.

Deane said at last, slowly, 'I started up this way from Sydney, months back. I was in trouble down there.'

'How?'

His voice was flat. 'The johns wanted me for a hi-jacking job at Randwick, and the track mob wanted me because they thought I had the lettuce. And I never had anything to do with it at all. I just travelled in the wrong company. It wasn't the first time—I've made a proper foul-up of myself. Maybe that's why I like to chuck my weight about when I see the chance. It blows me up a bit. The rest of the time I'm pricked through like a blasted toy balloon . . .'

The fat woman sighed. 'Better move on, Johnny. Move on, and try your luck again.'

Deane shook his head.

They were silent. Finally she said, 'What are you going to do?'

Desperation began to push him. His mind and his mood hardened as he recovered. He said, 'Forget it. Don't worry about motives. Forget all the other things. Malone killed that abo and I'll see him punished for it. Simple—that's the beginning and end of it.'

She sighed again. 'Malone's tougher than you think.'

'Just wait.'

'And what about the rest? And us in Galah?'

'Find your own answer, Margie. You'll have to, some time.' He moved past her to the door, walked out into the light and back down the main street.

7

He went round to the back of the courthouse. The afternoon sun was high in a white sky, blistering down. The leaves of the bottle-tree hung still. Nothing moved, nobody was in sight.

The backs of the houses, yards and privies, rubbish dump and tumbledown walls. A roaming dog poked through the pile of litter; and beyond was the bush. Bright sunlight killed in a brilliant glare, drought on the ground, the skeletal bleached framework of the fallen shacks under the sun like picked ribs against an empty desert . . .

Heat waves rose over the spinifex scrub, into the distance. Deane turned to look back at the courthouse above him, stone-built to last the years of sun and willie-willie; but now derelict.

He wondered if Malone's wife would come.

He kicked at the wooden door. It gave, grudgingly. He pushed harder and the door opened fully, scraping its path through the dry earth which had filtered in.

He stepped inside. There were the prints of children's bare feet in the sand, where the kids of the town had explored. The paw-marks of a cat . . . Deane walked into the courthouse.

He stood still, gazing up at the high ceiling. Light

came down from a hole in the roof, through broken windows, falling on the seats, bench, the polished wood.

Deane glanced back at the door, but there was no sign of the woman. He had no idea what time she would arrive; if at all. He moved across to the bench, through the dust and sand.

The court records were still there. He turned the covers, and the ink lay pale on white sheets. Crimes long past: the final entry was dated twenty-five years previously. He sat on the magistrate's chair and gazed out at the silent, dusty seats.

Then he shut the door, stretched out his legs. Everything was quiet. He closed his eyes, relaxing. His breathing made the only slight sound in the emptiness.

He woke abruptly. The sunlight had shifted an inch on the table. Standing at the end of the room by the door, watching his sprawled figure, was Malone's wife.

They looked at one another across the silence. In the high light her face was calm and contemplative. At last she began to walk forward, to the bench. Deane said softly, 'Well, hullo . . .' Suddenly he felt unsure.

She did not reply.

Deane pushed himself upright, opened again the pages of the dusty court-book. 'Enter. Come ye to be judged . . .'

She stood still, unsmiling.

Deane closed the book again. 'Glad you've turned up.'

'I didn't intend to.'

'Hubby know you've come?'

She remained silent.

He looked down, ran his fingers exploratively over the marks left on the table, traced through the dust. The white, gritty powder clung to his fingertips. At last he said thoughtfully, 'Dust. Of men, of life.'

She raised fine blonde eyebrows.

He said again, 'Dust. Always sombre and terrifying.'

'Is it?' She spoke, carefully.

'Dust falling, in·churches, in old houses. Like seeing your own name on an ancient tombstone. Like cold wind, the awareness of death. It sweeps through you.'

She lifted her eyes and looked at him with a changed expression, curiosity.

Deane shrugged it off. He said, 'How did you get here?'

'How d'you think? By truck.' She explained deliberately, 'I only travelled over to call on Grace Fisher.'

He smiled. Then he brushed down the bench beside him, wiping his dirty hands on his knees. 'Sit down.'

She stirred, stepped across and lowered herself to the painted wood. He studied her face: the structure of bone and flesh, eyes and lips. The line of her jaw had the slight fullness of maturity. She had an attractive, finely-made face; on it were marks of stress, gathered under the eyes.

He said, 'Now tell me your name.'

'Nancy.' She was unwilling.

Deane waved his hand round the falling, crumbling building. 'Fine setting.'

'You suggested it.'

'But I like it. I go for lonely places and silence. Even the dying air and forlornness.'

'You're welcome.' She sat in the shaft of light and it touched the fair, slight hairs of her face, her legs. She wore a white dress.

Deane said, 'Not your cup of tea, Nancy?'

She shook her head. 'Not mine.'

'I guess you didn't start off in this place.'

'I was born in Queensland.'

'And marriage brought you here?'

She was silent. Then she said, 'Yes.'

'You got a good rich station to live on, anyway.'

She didn't move. Then she looked up, her gaze passing over him.

Deane sat back. He said, 'They tell me the drought makes people do strange things, around this quarter.'

'It kills you.' She held her eyes against his. 'It drives you crazy. Makes you do things you'd never think of doing otherwise . . .'

Deane said slowly, 'I guess it does.' Reluctantly he began to feel inside him appreciation for her, a softness. She had lowered her head. Something in that white, bent neck beneath the short-cut fair hair touched him again. Suddenly they were close once more, but he knew sympathy instead of desire.

She said, 'Heat and death and loneliness. They kill. Oh, they kill . . .'

'I know.' Deane glanced across the shadows of the court, up to the ceiling. Then he said, 'I'm sorry.'

She shook her head. 'Maybe I'm glad after all. Sex is life. Better than the paddocks, dying cattle in the sun . . .'

They were silent. There was the sound of a barking

dog from the street beyond. Deane changed his position, stretching out on the seat.

He said, 'I had a bit of an argument with your husband this morning.'

She raised her head and turned to look at him. 'I heard about it . . . What's the trouble between you?'

'Maybe he's found out . . .'

She turned away.

Deane said, 'Seriously. Don't you know?'

'No. Of course I don't.' She was disturbed.

Deane said, 'He killed an aborigine who worked on your station. Thomas Clancy . . . He hung the poor blighter up from a mulga tree. I don't see that he should get away with it . . .'

She stared at him. 'Thomas . . . ?'

'Trouble about a black girl.'

She said, 'I've been hurt too often before . . .' Her own voice was bitter. Then she asked, 'What's it got to do with you?'

'You're as indifferent as the rest about it.' Deane spoke harshly.

'Indifferent . . .' She stopped. Then she burst out. 'No. Of course I'm not indifferent. I know the way he's behaved, some of the things he's done. I care—God, I care. But what's the good? What's the stupid, crying good?' She got to her feet, stood rigid beside the seat.

Deane said, 'All right, Nancy. All right.'

She sat down slowly and looked at him. 'You're an unusual bloke. You're a strange one. English, aren't you?'

'Just another pommy,' he said.

'You're a mixture.' She regarded him with curious intentness which momentarily absorbed every other emotion. 'You change. Sometimes you're brutal—and yet you feel about things. Sometimes you seem to talk rough deliberately—and at other times you've the real accent. You're strange . . .'

Deane got slowly to his feet beside her. They stood together. She held forward her hands for an instant, as though she were about to push him away. She had long-fingered slender hands, and two rings glittered on them. Then her arm moved and, instead of pushing him away, the hands settled on his side, holding him. Deane put his own arms round her and he felt the firm flesh of her back under his fingers, quivering through the thin dress.

They stood among the stirred dust, the silence. He felt tenderness, warmth for her suddenly; he recoiled, wrenching his eyes from her fair and flushed face in the light. He said sharply, 'Time for me to go, Nancy.'

She stepped aside. She looked up at him in perplexity and then she said quickly, 'I was leaving anyway.'

'Maybe I'll see you again.'

She grabbed her bag and walked away, passing down the corridor, outside. He heard the rear door swing and grate behind her.

Everything was silent again. Deane cursed, and then he followed her out.

He saw her away down the street, hurrying towards a Land-Rover parked near the store. The bright sun wrapped her figure, flamed on her fair hair as she walked erectly. Once again he felt softness stirring in him, unsettling him like weakness in the stomach.

He strode away rapidly in the opposite direction. He wanted no claims, no bonds. He walked savagely, stamping his booted feet against the baked earth of the square. Emotion unsettled him: indifference brought calm. He went into the hotel, across the wooden veranda.

He walked down the corridor in ill-temper. Then near the hallway he met the owner's wife.

She smiled readily. 'Good day, Mr. Deane.'

Deane nodded, brushing by.

'Getting fed up with us yet?' She smiled again, a gleam of large teeth.

He stopped, looking back at her. Suddenly attention came to him. He feared any burden of tenderness towards Malone's wife: but this was not the chance of killing it. He looked at the hotel woman again, her thick waist and big bosom, her ripe, lush face.

He said abruptly, 'Not yet, Grace. Nowhere nearly.' He grinned suddenly down on her. His grin was mocking, deriding her, himself.

She said, 'Now you don't expect me to believe that.' She was smiling back instinctively, aroused and willing.

Nobody was about. The door of the room was open behind her and he saw a large double bed with pink covers. Doubt flickered over her face for just a moment, and then it gave way to anticipation. Her tongue passed across her lips. She said, 'Want to see in there?'

'Some other time.'

He spoke roughly, turning away from her astounded, angry face. He swung on his heel and went back outside, leaning against the veranda post, gazing over the square

93

and breathing the hot still air. All on his bloody own. And the only burden on his back was himself, and his own stinking contempt for himself.

Then he saw the policeman, passing by down the street, in his trim khaki attire. Deane felt restlessly spiteful. He shouted out. 'Time you took a trip around your beat, isn't it, copper? You don't seem anxious to step far off your own cushy doorstep . . .'

Lawrence swung round. His eyes flashed, and then he came over to the veranda. Anger splotched his face. 'You take it careful, sport. Just take it careful. You'll be locked up if you don't watch it, and you'll stay there. You'll stay there as long as I say.'

'Get away. You don't want me around Galah a day longer than you can help. You don't want to hold me— you want to kick me out sharp. Don't you now?' Deane jeered.

Lawrence took a step up to the veranda, stopped. His throat jerked. 'So you think so? Just tell me why I should want that?'

'Because I know what your blasted friend Malone has been up to. And because I'm not frightened of what he can do, in the way the rest of the town is. Because I don't care a kangaroo's tit about this lousy hole called Galah, and I'm ready to see it tumble rather than let Malone get away with anything. Because . . .'

'Shurrup!' Lawrence shouted at him. His voice echoed under the eaves. He took another step forward. Anger burned in the brown, tight face. 'All right, pommy, all right. So you're not in a hurry to go. Then watch yourself. Because you're going to be unpopular, and I'll not come running when you call. You're on your own . . .'

'That's the way I like to be.'

'Then enjoy it.' The policeman pulled on his hat-brim and turned away.

Deane sobered down. He relaxed, and leaned against the post. Then the copper swung round and came back to him.

'Something I forgot to tell you.'

'Well?' He spoke tiredly.

'I don't trust you. Smell something funny about you —and when I do that I'm pretty often right . . . I've started off some inquiries about you, sport. Way back down the road, and things don't stop there. In the Territory or outside it, that's not going to matter—if there's anything against you, I'll get it through. I'll have you checked through every police station in the country . . .'

'And the best of luck.'

Lawrence said, 'You should've done what I suggested. My job is to look after Galah. I'm going to see to just that. You've had it now, pommy, all down the line . . .' Then he walked away.

Deane sagged against the post again. He felt indifferent.

He stared at the bare earth in the sun. After a while he heard a low whisper. 'Boss. Aw, boss . . .'

He looked across. The old aborigine named Albert squatted on the edge of the veranda near by, watching him. Deane said without enthusiasm, 'Well?'

The abo pulled his dirty coat around him and shifted his bony haunches a few inches nearer. 'Boss—you tak 'im longa Clancy Rock, dis feller? We go see dat feller Jack, eh?'

Deane turned away. 'Not now, Albert—some other time. Me fixem, some other time. Longa next day, maybe.'

'Him crook, boss. Him bin needem doc.' The abo's voice was urgent.

Deane said sharply, 'I'll see you later, Albert. I'll let you know.' He stepped down from the veranda and walked off down the street, off to the edge of the township. He stared over the bush, at the stony trail which led out. Away north, to the Anderson homestead . . .

He went across to the derelict shed where he had left his Austin truck. He checked the engine, wiped it down, started her up. She ran easily. He sat on the leather seat and stared again at the horizon.

It would be good to drive out and leave everything behind him, the way he always did . . .

The young children who played in the streets came clustering round at the revving of the engine, regarding him warily. They had dark skins, muddy skins, just a few white; they had aboriginal features and European features, intermixed. He got down from the truck and they scattered hastily, watching from a respectful distance. He wondered what they had been told about him. He stood and gazed around him at the broken walls and empty windows, and the closed shacks of an unfriendly, antagonistic town. Stranger in Galah, unwanted.

Well, he asked for it.

Later he went back to the hotel for the evening meal. He strolled outside afterwards and sat on the veranda rails in the cooling air, legs dangling over the rails, lonely as ever. The sun slanted over the roofs and the sky turned to pink, to red. The evening stretched long.

An old American car rattled down the main street and pulled up outside the hotel. Deane eyed it: a Chevrolet. A young man got out, reached back into the rear seat for a long parcel. He glanced at the hotel and then, carrying the package as though it were heavy, he climbed up to the veranda.

When he was near Deane knew his face; then he recognized him as the final remaining participant in that first encounter among the bush, whom he had come to meet again: the young white man who had sat in the back of the Land-Rover and had covered with a rifle the captured aborigines. Who had helped Malone to murder . . .

He regarded Deane; he was somewhere in his early twenties, his face still boyish and clean-lined. He wore cowboy shirt and belted dungaree trousers, an Ashburton hat and elastic-sided boots. He moved lithely, but his face was full of enmity.

He flung down the parcel on the wood floor of the veranda. 'Got a present for you.'

Deane said, 'What is it?'

'Open up and you'll find out.' His voice was hard as his face.

Deane shrugged. He got off the veranda rail, pulled the wrappings away. Inside were a pair of truck springs. He checked the parts label and they were of the correct pattern for his Austin.

He straightened up, cautiously. 'Where are these from?'

97

The young man laughed shortly. 'Clancy Rock.'

'I thought the mail plane wasn't due until Friday.'

'Malone wanted to do you a big favour. He got me to fly his own private aircraft down today and fetch them back specially. Four-hundred-mile trip, just for you. Hope you're grateful.'

Deane stared at the springs.

'Malone's proper anxious to help you. But he sent a word of advice.'

'What's that?'

The young man said, 'He suggests you get them fitted bloody quick. And when that's done you don't waste any time in using them. It's a long road north.'

'All right,' Deane said. 'All right.'

The young man stared at him with the same hardness. Then he swung himself over the rail, jumped back into the Chev and drove away. The red dust scattered.

Deane stood looking at the black springs. Then he gazed over the street at his parked truck. Finally he picked up the springs and carried them across.

He jacked up the body of the Austin, supported it on piled lumps of stone. He stripped down the existing suspension, the broken leaves.

Then he stared up at the sky. He told himself it would soon be dark, and there was no time now to fit the new parts. He piled the springs into the cab of the truck and went back to clean himself up, leaving the job unfinished.

When he came out the twilight was settling down fast. The main street was empty. Deane stood watching night fall over the silent roofs, the few lights gleam on. The stars began to sparkle in a glittering sky.

98

The Czech waitress came outside from the back of the hotel, walked round to the front near the steps. She had taken off her apron and she was neat and small in the black dress, against the night.

Deane watched her unemotionally for a while. He shifted his position on the veranda and the planking creaked. She turned round, startled, and glanced up at him in the dim light of the porch lamp.

She smiled. 'Good evening.'

''Evening,' he said.

'It's a lovely night. Cool, after the heat.' Her voice was soft.

Deane shrugged his shoulders without comment. He had had enough of women for one day.

She glanced away.

But she seemed more like a girl, friendly and simple, a wisp of a figure in the lamplight. He said abruptly, making conversation, 'What do you do around here on your nights off?'

She was smiling once more. 'There is nothing much. I look at the stars . . .'

He strolled down the steps, leaning against the wall beside her. 'Worth looking at. But dull?'

'I don't mind.'

Deane grinned. 'You seem to take things pretty quietly.' He spoke to her as one way of filling in the time; no likelihood of getting himself involved here.

'I like peace.' She was calm. 'Once I missed that.'

He glanced at her for a moment, then was silent. She was a foot shorter than he, little more than a shadow beside him in the dark; but her face gleamed clear, intense.

'You've got some courage,' he said suddenly.

'Why?' She was puzzled.

'To choose a life and district like this.'

'Choose?' She said, 'One does not choose. One accepts. And there's no courage in acceptance.' Then she pointed to the Southern Cross. 'How near tonight. How real.'

They stared at the sky. Deane turned aside. 'It's getting cold.' He hesitated. 'Come for a drink in the café.'

She looked at him, startled. Then she said, 'Yes. I would love to.'

They walked down the street, into the yellow light of the empty café. The radio played dance music. Deane said, 'Take a seat.'

She slipped into one of the wooden cubicles and he went over to the counter, got two cups of tea from the half-asleep Greek proprietor.

He took them back to the cubicle. 'How the heck do these people make a living?'

She lifted slight shoulders. 'I think they must be satisfied with very little . . .'

Deane dropped into the seat opposite her. She lifted the cup of hot tea with small, pink-nailed hands, looked at him gravely over the rim.

He said, 'Well? Any comments?'

'I have been learning about you.' She had black, straight hair above an open forehead, eyes which were vivid and violet.

'Who from?'

'Mrs. Fisher—and her husband.'

Deane's eyes dropped. At last he said, 'And what have you heard?'

'That you are not very popular in this town, and the reason why . . . But Mrs. Fisher thinks you are quite the—the gentleman.' She was serious.

'So she does?' Deane was briefly amused. 'Surprising . . .' He looked down over himself again, the sweaty old clothes that wrapped him in, stinking to remind him of all he had come to be . . . Then he asked, 'And what do *you* think about it?'

'I admire you.'

Deane pushed away the teacup. 'Hell,' he said. 'Not that . . .'

'For asking questions about the death of that black man. For not just ignoring it, the way the others would do.'

'You don't know my reasons. They're not good ones.' He spoke roughly.

The foreign girl took a pack of cigarettes from her bag, offered him one. Deane shook his head. She lit, using her own match, drew in. With the smoke curling about her face she looked older. She shook her head quietly, watching the red tip glow; then she said, 'I have been here long enough to know how this town is, and why they are unfriendly to you. And I know Richard Malone. But these others are not bad people—they have their lives to live.'

'They're welcome.'

'I know it is harder for them to have a conscience, the way you have.'

Deane was stirred by sudden violence. 'Conscience?

Are you crazy? You don't know me, girl. I'm selfish as they come. I've done some pretty lousy, pretty dirty things, the few days I've been here. Now my truck springs have come and tomorrow I'm on my way. I shan't leave anything good behind me, I'll be no loss. Nothing to admire about me, Elsa . . .'

'You're going tomorrow?'

'Yes.'

She nodded slowly.

Deane said, 'I'll be moving on. And Malone can drive round stringing up every abo in sight for all I care.'

'Where are you going?'

'Nowhere. Anywhere. I'll find out.'

She was silent. The radio drifted into crackling, picked up as the static cleared. She said again, 'You mustn't blame this town. It is only the drought, and the fear. Fear of the passing of Galah if Richard Malone should fall, and Clancy Rock Station with him . . .'

'I've heard that before,' Deane said. 'It doesn't convince . . . Anyway, it's nothing to me now. This town means nothing.'

She looked up at him. Her eyes met his through the swirl of cigarette smoke, across the pale light of the café. They were woman's eyes, reminding him of the eyes of Malone's wife, calm and deep; and he knew this was no girl but a woman too, for all her slender body.

He said, 'Let's go.'

'Already?'

'Yes.'

She stood up. He followed her out into the street. They walked back to the hotel without speaking.

At the foot of the veranda steps they stopped. A faint night wind blew over the street. The roadway was empty. Deane said abruptly, 'I told you I wasn't much to admire. And that isn't self-pity—just a straight fact.'

'It's all right.' She stood looking gravely at him. 'I hope you have a good journey tomorrow.'

'We'll see.' Her steady eyes were set in a sober face, a dark figure in the black dress among the murky shadows. He said, suddenly more gentle, 'Good night, Elsa.' Then he walked away, strolling around the silent streets of the town again before he came back to the hotel with its faded sign, the lamp gleaming over the dark square.

He stood in his room, staring back across the street to where his truck was jacked up. Tomorrow he would finish the repair: and then pull out and run, the way he always ran.

8

HE started work on the truck again in the early light. Sweating and heaving, he fitted the new springs, replaced the wheels.

Then he sat still, feeling the wetness of his shirt against his back. The warmth of the sun spread. Galah Creek was lifeless, surrounded by the miles of bush to the pale horizon.

He went over to the hotel to fetch his bags. In the hall-way he flipped money at the bleary-eyed hotel manager, strode down the steps and crossed back to the truck. He packed the food, cans of water and petrol, stowed everything away. Then he went to the driving-seat and got in.

He was all ready to go; but he sat there for quite a while.

Then through the spotted windscreen Deane saw the policeman leave his office and come across the empty street towards him. He waited.

The copper leaned his hand against the door frame and stared in. 'So you're off?'

'Some guess. You policemen are wonderful.'

The policeman's cracked lips tightened. 'All right, sport. All right. I'll let that go.' He spoke slowly, forcing out the words. He hated Deane the more because underneath he felt guilty. And a ragged, shiftless down-and-out should be aware of his guilt . . .

Deane grinned back knowingly.

The policeman pulled himself upright, straightening his back. He stared over the trail. 'Get moving then.'

'It's a pleasure.' Deane started up, and the engine clattered in the silence with a suddenness which would have woken the birds in the gums. But there were no galahs left to fly . . .

Abruptly the policeman said, 'Just a minute.'

Deane took his hand off the gear-shift.

'Well?'

Lawrence hesitated. At last he said, 'Listen. I'm going to make an offer to you.'

'What sort of offer?'

The policeman's faded blue eyes looked at him, then skidded away to the empty street. He wasn't sure how to put it. Then he said, 'You've got a past, pommy. Don't fool me about that—I know you. You've gone against the law, somewhere back down the trail. And I'll get it through, sooner or later . . .'

'Think so?'

'Yes.' Lawrence said, 'But once you flit out of this place, I'll have no hard feelings. My territory's all I'm concerned with. No reason for me to pass the tale on after you.'

'Nice of you.' Deane began to smile knowingly again.

The policeman's eyes moved back to him for a moment, then lingered on the dull paintwork of the truck bonnet. 'So you forget us when you move out, the way I'll forget you. Eh?'

Deane said, 'Fair exchange . . . But just supposing you're wrong about me, and I'm pure and innocent as a newborn lamb . . . Not much of an offer then, copper.'

'I'm not wrong.' He was sharp with resentment at having to deal with Deane. His lined face was stiff and his fists were tightened. He hated it. 'I'll stake my life I'm not wrong.'

'Never gamble, copper,' Deane said. 'We'll see. We'll just see . . .'

Then he let the clutch in suddenly and moved off. The policeman jerked himself aside, staggering. The Austin accelerated down the street; Deane glanced behind in the mirror and saw Lawrence standing in the road, bare-headed, watching him go. Once more the mocking grin spread over Deane's face.

He brushed it away, pulled down his hat, concentrated on the dusty road ahead. The petrol and water splashed in the cans behind him, and he drove out of Galah just the way he had driven in.

Except that there was no limp, fly-crawling abo's body sprawling in the back of the truck.

That set him off thinking, and he didn't enjoy it. And then he felt guilty as the copper had done, and there was the same reluctant stiffness on his own face as he went, running out.

He drove slowly. Mulga scrub and spinifex, and low hills on the horizon: the sun came up and the heat thickened. The road stretched flat and straight to the north, past the next homestead and then on.

Galah was gone. For a while he felt relief at being away: freedom again, solitude. A man and a truck, alone in a vast country, unknown, stranger . . . Deane opened his mouth to sing raucously. But the gritty dust got in, and he was silent once more.

The truck rolled forward, rising and falling, battering on down the rough road. A few miles farther on Deane stopped for a swig of water, a bite of breakfast. He pulled up under a clump of mulga trees.

He sat still under them, munching, gazing out over the empty, hazed bush. He looked up at the tree above him, the grey-green, dead and brittle wood of the branches. Suddenly he remembered how another tree had crashed down carrying the body of the abo, the noise of its falling and the ugly face under the brushwood.

It flickered through his mind, mixing with the face of the second aborigine who lay in the dark cavern with bullet holes in his stomach, and the starlighted barren plain below. And he saw the face of Malone's wife, and the red eyes of Malone, the grave eyes of the Czech girl. 'I admire you . . .' The open street of Galah and the shacks; and then again that first moment in the bush when he had bent over the fleeing aborigine, seen the terror and pleading in black eyes . . .

Impatiently he kicked out, and the dead, ant-eaten mulga cracked under his boot. Slowly he got back in the truck, engaged gear and travelled on.

He went another couple of hundred yards. Then he took his foot off the pedal and let the truck come to a halt. He sat still while the engine ticked over in front of him, alone under the sun and sky, the silence and emptiness. He did not stir. Then he cursed bitterly. Then, moving heavily and driven almost against his will, he began to turn the Austin round.

The tyres slashed at the dust on the edge of the track as he reversed. The truck straightened out and started back, travelling very slowly, moving back along the way he had come.

He sat over the wheel, hunched down, sunk in himself. Several times again Deane's foot lifted on the accelerator, shifted to the brake, hovered . . . But each time he drove on. The still bush journeyed by, the sparse mulga trees threw brief shadows across the truck as it went.

A couple of miles out of the town he stopped, lifted the bonnet and disconnected two spark-plug leads. Then he went on, foot down, with the engine lurching and rocking before him like a sick thing.

He came back into Galah at noon. He pulled himself up as he came in, thrusting over himself the same roughness, the same hard style. He wiped his face dry and began to grin, anticipating.

The main street was empty. The sun was at zenith and the town lay flat and formless under it. He parked the Austin in the spot he had left it before, near the police station. Then he got out and carried his baggage over to the hotel.

He strode down the corridor, dumped the grips on the same bunk. Nobody was about. He went outside again, and the first person he saw was the policeman.

Lawrence had come out of his office and he was standing with his back to Deane, gazing fixedly at the truck. A few children gathered about. The post-office man leaned over the rail in front of his shack. Deane walked across.

He said, 'Back again, copper.'

Lawrence swung round. 'What's the flaming idea?' His voice burst out fiercely.

Deane shrugged his shoulders, smiling. 'Trouble again—engine packed up.'

Lawrence's anger raged over him. 'What sort of trouble?' His expression was wilder than Deane had seen it, fury and anxiety joined.

'Missing on a couple of cylinders.' He reached into the cab, started her up.

The policeman turned his head to the lurching tick-over. Then the engine stalled. He said, 'Let's see it.'

Deane shook his head. 'I'll handle my own repairs.'

Lawrence peered into the cab. 'You've travelled over sixty miles. You must've passed by the Andersons' place.'

'How do you know what I've done?'

'I checked your speedo beforehand.' He stared uneasily at Deane.

Deane said, 'I thought I'd rather come all the way back here. Nice, friendly town.'

Lawrence's gaze raked over him. Then he snapped, 'Turn around and get out.'

'Not bloody likely.'

'Go on.'

'No.'

Lawrence waited, tensed up. At last he said more quietly, 'How long will it take to fix?'

'Don't know.' Deane said, 'Could be quite a while.'

Lawrence swivelled on his heel and paced back to the police station. Deane locked up the truck and carried the rest of his gear to the hotel.

He met Fisher's wife. Her big face altered to surprise, then anger, then reluctant eagerness. She said, 'I thought you'd gone . . .'

Deane grinned again. 'Not yet, Grace. You'll see some more of me yet . . .'

He passed by, stacked his bags beside the bunk. The hotel-keeper poked his lopsided face through the doorway. 'You again.' Fisher's tone roughened. 'We don't want you, mister. Why don't you get to hell out of it?'

Deane looked back at the dirty-vested figure, the bony shoulders. He was going to snarl something pretty rude, but then he said mildly enough, 'I pay my bills promptly. Now push off.'

The man came forward into the room. 'You better be the one to push off. You better . . .'

Deane straightened up. He said, 'Finish it.'

The hotel-keeper paused, still cursing, and then he turned and went away, padding down the corridor in loose shoes.

Deane walked along to the dining-room. The girl Elsa was there. He said, 'Got anything for me?'

She looked taken aback; then she flushed slightly over the pale skin of her cheeks. 'I will see.'

He sat down at a table.

She came back from the kitchen. 'There isn't much. We weren't expecting—'

'I know.'

She put some beef in front of him.

He said, 'Did you have to fight the cook for it?'

She laughed.

'Anyway, thanks.'

The girl said, 'You changed your mind?' Her voice was quick.

Deane hesitated. 'I'm back for just a while.' He began eating.

When he went out the girl stood at the other end of the room, near the kitchen door, watching him. Deane raised a hand in farewell. She smiled back immediately, her cheeks flushing again.

He walked down the street and located the old aborigine in his usual spot against the corner-post. He said, 'Hullo, Albert.'

The aborigine stared up at him. Then he said, ''Lo, boss.'

Deane squatted beside him, glancing around the street. 'What's to do?'

The abo's gaze was thoughtful.

Deane said, 'That feller Jack Clancy—him still all right?'

The aborigine turned up a greasy coat-collar. 'Dunno, boss, dunno.'

'Him still longa same place?'

The old boy was motionless. Then he nodded.

Deane waited for a moment; he suggested. 'We go see 'im, eh?'

The aborigine's eyes did not stir. He was as cautious as Deane. After a while he nodded again. 'Longa what time, boss?'

'Right now.' Deane spoke more rapidly. Nobody was in sight down the road.

The aborigine relaxed. 'Awright, boss. You gottem truck?'

'We'll take mine.'

Deane walked back to the Austin. He raised the bonnet and refixed the ignition leads. Then he drove across the square. The aborigine got to his feet and hobbled on board. The truck moved forward down the street.

He knew eyes were watching them from the shacks. The old abo turned to him. 'You best watchem out, boss. Dese people no likem. You bin in plenty trouble now, boss.'

'I'll worry.' The truck started out on the road for the station.

The aborigine sat staring from the opposite window. The bush passed by. Behind them nothing and nobody was in sight. But their tyres left marks which would lead a tracker for miles, over rock and creek-bed, over saltpan and flat. If Malone wanted Jack Clancy, he would find him pretty soon . . .

Deane turned to the aborigine. 'This feller Jack, maybe we better shift 'im. Find 'im some other place longa bush, eh?'

Albert nodded. 'Yes, boss. We shift 'im.'

'Where?'

'Me know place.'

They turned off the main track. They were alone to every horizon. Then the ground became stonier, burning hot in the light. The rock appeared ahead, gleaming against the sky. The Austin crunched over the stones, drew round the back of the outcrop and pulled up.

They climbed to the cave. Sunlight struck in, but the far corner was in gloom. The wounded aborigine lay in the same spot. The can of water was empty but the kangaroo meat was hardly touched. He lifted his eyelids wearily as they came across. He had weakened.

Deane bent over and examined him. Sour waves of smell attacked his nostrils. The stomach wounds didn't show any external signs for the worse, but Deane was no expert. Anyway, the abo was still alive and still conscious. Deane breathed out, relieved of a burden.

He said, 'Time we moved him—too chancy here. Which way?'

The old aborigine pointed over the stone flats, way below. 'We tak' 'im in truck, boss.'

'Let's go.' They lifted the thin body, carried it down the slope of the rock. The wounded abo did not speak or moan. They put him in the back of the truck and Deane drove away slowly. 'Where?'

Albert pointed a dirty finger.

Deane said, 'Bimeby, we tak 'im into town. Him speak to that feller policeman Lawrence, tell 'im 'bout Malone. But not just yet.'

The abo looked uneasy. 'No good, boss. That no good. Feller policeman, he tak 'im no notice.'

'I'll see to that.'

The abo said, 'There, boss. Over there. We catch 'im now. We stop.'

Deane pulled up. They lifted out the wounded aborigine and carried him to a small outcrop of rock among several others which dotted the land, on the opposite side of Clancy Rock to the station. They hid him in a crevice a few feet up from the ground, a cavern similar to his previous hiding-place. They left another can of water, some beef.

The two aborigines muttered together. Then the injured one lay quiet. He was too sick to care much about anything. Deane and Albert came away, walked to the truck. The vehicle started back towards the main road: over the hard ground their tyre marks would be difficult to follow.

The truck headed across country, picked up the trail and drove towards Galah.

The old abo said, 'Him not last long time now, boss.'

Deane nodded slowly. 'We'll fetch him away soon. We fetch 'im, Albert.'

He dropped the abo on the edge of town, drove into the square and parked. Stiffly he got out of the truck. It was late afternoon. The policeman stood waiting for him. 'So you're a liar, pommy?'

Deane said, 'Eh?'

'Nothing wrong with your truck.'

'So there isn't . . . I cured it.'

'And now?'

'Well . . .' Deane said, 'I'm not in any rush to go, after all. Not for a day or two.'

Lawrence watched him. Then he asked, 'Where have you been?'

'Out for another ride.'

'I don't like the company you keep.'

'What's wrong with it?'

The policeman paused. The hotel-keeper was standing at the bar door. Beside him was the green-shirted cattle-hand who had fought with Deane before: he had a gashed cheek. A couple of aborigines sat in the dust to watch. The café proprietor had come outside. A couple of other townsfolk were gathered close, two men. Finally Lawrence said, 'You're going against this whole town now, pommy. You're trying to pull it all down, and we're all against you. You're asking for a lot of trouble, and soon I'll be doing more than threatening.' He spoke heavily. 'You've got no chance left to win, sport.'

Deane stood still, glancing around and counting the odds. 'I've nothing against the town.'

'When you join with the abos, you're against us. You're attacking us all.'

Deane kicked at the loose soil. He was no damned hero. 'I'm not joining with anyone. I'm just seeing a thing through.'

Lawrence came nearer to him. Faint grey bristles

poked through on his chin. 'What thing? Just what do you want?'

Deane raised his voice abruptly. 'You know what I want. I want you to arrest Malone and send him down for trial on a murder charge. When that's done I'll drive out.'

The policeman's gaze became still harder. He stared round the square, then at Deane again. He was disturbed. 'What's Malone got to do with you? What've you got against him?'

Deane shook his head. 'Just say that I've run far enough, copper. Here I dig in. You'd better hold Malone.'

Lawrence's face set. 'You're crazy. You drifting, foot-loose hobo. You—you—telling me what I'm supposed to do.' Deane had got him unsteady. Again he glanced round at the onlookers. 'Attacking a bloke like Malone. The man who makes this godforsaken district, who holds it together . . . You give me the guts-ache, you floggin' liar . . .'

'Arrest him.'

The copper swallowed. 'I'll arrest you first.'

'Arrest him.'

He pressed nearer. 'You're asking for it. You're asking . . .'

Deane jeered. 'You don't dare.'

Lawrence stopped abruptly. 'All right. I could jail you but we don't want the trouble of keeping you.' He swallowed again. 'You'd better be out of Galah by sundown tonight. Be out, pommy, that's a warning. Because if you're still here you'll deserve all you get.'

'And what's that?'

'You'll discover.'

'Come on then.'

The policeman stood rigid. He fixed his hard gaze on Deane, trying to stare him out. Deane's eyes did not shift. Finally Lawrence pushed past him and walked on.

Deane was left there. He cooled down fast, facing the opposing group of hostile faces, the hotel-owner, the post-office man and the café proprietor, the cattle-hands . . .

The mail-office man stepped over to him. He was bony and fleshless like the hotel-owner; his bald head was sweaty in the sun. When he spoke a few teeth gaped in his mouth. 'Who are you to judge Malone? Eh? Think you're any better?'

Deane said, 'I'm not judging. I'm accusing. Some-one else can act the judge.'

'Think you're any better, I asked.'

'I don't go around murdering abos.'

The post-office man swore. 'You creeping lousy no-good. You think we're going to stand for you . . .'

The others moved nearer. They were fierce, menacing him. And underneath their expressions that flicker of disquiet because they knew that Malone was a killer after all . . .

The green-shirted cattle-hand was in front. Deane said, 'Going to take me on again?'

The cattle-hand came towards him. Deane got ready to fight. Then the hand said, 'I'll bring a few other jokers from the station with me. No fightin' straight with a cow like you. We'll make a job of you later . . .'

He turned aside, went back to the bar. Deane dropped his hands and grinned tightly at the rest. He was ready to take them all on. But they changed their minds. They cursed at him, and split up . . .

They went past, ignoring him. The street began to empty. He noticed standing there the fat woman from the store; he hadn't seen her before. Deane crossed over. But she stepped aside. 'Keep away from me, Johnny.'

Anger swept him. 'To hell with it, I won't.'

She began to walk back to the store. He went beside her. 'Go away,' she said.

'Blast you.' He gripped her arm. 'What've I done? Nothing so bloody vile. Accused a man when we all know he did it. What's that against me?'

She stopped and looked up at him. 'You've no right to accuse, Johnny.'

'Sure I've the right. I'm not unclean, I'm not crooked. I've done some rotten things but I've still the right to jump on a killer when I see one. Murder's the greatest crime.'

'Not always,' she said. 'Not always by half.'

Deane grabbed her arm again. 'Listen, Margie. For God's sake listen to me.' His voice was suddenly unsteady. 'This gets me down—you at least ought to understand. It's a long while since I stepped out of my way to help anyone—hell, it's time I did it now . . .' He was walking quickly beside her, stumbling through the sand.

She threw his arm off. 'Help? Who d'you think you're going to help? You're only goin' to hurt again, to destroy. The abo's dead and the rest of us here can only get damaged by you, you dirty, ill-tempered louse—'

'Blast you. Listen—'

'I'm not going to listen.'

'The abo's dead. But the way Malone is, plenty other abos could follow him. And not only blacks. He's up the pole—'

She turned on him. 'Get away. Just get off, damn you. I don't want to talk to you. Get going . . .'

Deane stopped. She pulled herself away, waddled fiercely across to the wooden sidewalk of her store, disappeared through the door.

He stood in the dusty road. A group of children in tattered dresses screamed at him. He was motionless for a while and then he began to walk slowly to the edge of the town, to the edge of the infinite bush. He could have got away, and he had come back. He stood there, with the sound of their yelling behind him.

9

THERE was the noise of a car plunging across the lumpy track behind him, braking. Then a rough shout. 'You!'

Deane turned slowly. The young man who had brought over the truck springs from Clancy Rock stood a few yards beside the battered Chevrolet, his hands on his hips.

Deane eyed him with bitter distaste.

'Got a message for you from Malone.'

'Well?'

'Come over here and I'll give it you.'

Deane laughed harshly, then walked to where he stood. 'Well?'

'Malone wants to see you.' He had a youthful, handsome face, dark hair; he was hostile.

'What about?'

'Don't know.'

'When?'

'Now.'

Deane laughed again. 'He's cutting it a bit fine. Doesn't he know the copper has warned me out of town by sundown? Won't give me much time for social calls before I go . . .'

The young man stepped forward. 'You're leaving now?'

Deane ran a thoughtful hand over his chin. 'Well . . . You never know about that . . .'

'You better come.' The young man jerked a thumb at the Chevrolet. 'Get in.'

Deane paused; then he shrugged. He could take care of himself.

They stepped into the car. He sat on the front bench seat beside the youngster. The Chev's big tyres flung up the dust behind them and they swung out on the road.

The car swayed over the potholes. Deane said, 'Crowd of you in Galah today. Wonder what brings you all here?'

The young man didn't reply.

'Can't be you've come to welcome me back.' Deane relaxed, leaned against the seat. Then he said indifferently, 'Who are *you*, anyway?'

The young man wrenched the wheel. 'Mind your own damn' business.'

'Oh, stop playing the tough feller.' Deane's voice was tired. 'You've done enough of that.'

The young man glared at him. Then he stamped down on the accelerator and the car bucked and plunged over the road surface, throwing them about in the seat.

Finally the youngster said, 'If you want to know, the name's Ted Gair.' He slackened the pace of the car.

'What's your job at the station?'

He hesitated again. 'Jackaroo.'

'You've learned some fine tricks from Malone. Lesson one—how to kill abos, heh? Take a long rope . . .'

'Shut up about Malone. Think you could run a station this size? He's a big man . . .' His voice was uneasy.

'He'll fall,' Deane said.

The young man glanced at him again. 'Sometimes you have to do these things . . .' He speeded up the Chev once more, and they came to the fencing-wire. He ordered, 'Open the gate.'

Deane said, 'Open your own bloody gates. I'm the guest on this trip.'

Gair got down and threw back the gate, drove through. Inside the paddock he stopped, walked back and closed the gate behind him. He strode across the rough ground in the evening sun, his face furious and yet unsure. Deane grinned at him as he got back into the car. Gair drove on, wiping his hand across his mouth. He didn't seem certain how to behave.

They travelled over the stillness of the paddock. Shadows were longer. Malone's cattle stood in heaps across the land. They waited as they had waited before, for the ending of death. Their eyes did not shift as the car drove by, their heads did not lift. A cloud of flies buzzed, buzzed. The land was vast and empty and the ground was bare. The sky was hot.

Deane said, 'Gives me a thirst.'

'Gives me the creeps.' The young man shuddered.

'Not used to it yet?'

'First drought I ever saw like this one. It's not so pretty.'

'Don't you like to see death, Ted?'

Gair swung his head quickly away. 'Dry up. Do the rest of your talking to Malone.'

Deane stared out of the window. The car rattled on towards the station. Away to the horizon he saw the high, ragged outline of Clancy Rock itself.

He began wondering what Malone would have to say to him . . .

He caught a glimpse of himself in the twisted rear mirror, and disliked it the way he always did: he saw a rough and ragged-looking bastard. Suddenly he wondered why any of them should trouble about him, because his threats could carry no more weight than his promises. Who was going to listen to him, derelict pommy-on-the-run, when he spoke of justice? Fornicating pommy-on-the-run, on the bum, on the way down . . .

He slouched low in the seat with his hat over his eyes, watching the buildings of the station come nearer over the paddock.

They pulled up near the homestead. Gair swung his legs out. 'All right. Come on.'

Deane followed him across the yard. The couch-grass lawn had yellowed and the flowers were beginning to wilt. Even here the drought was tightening its hold.

The young man led the way into the office. It was a shed near the store, an outer room and then a section with a desk, a couple of chairs, paper files. A map of Australia and an enlarged section of the Northern Territory were fixed on the bare walls. The windows looked out on the central area, bunk-house, the homestead beyond.

Gair said, 'You better wait here. I'll fetch Malone.' He glanced round the office.

Deane took a chair and stretched out his legs. 'I won't pinch anything.'

'Don't try it.' The young man stood in the doorway. He regarded Deane with uncertainty on his frowning face, as though he didn't know where to place him. Then he went out, over the yard.

The station was quiet. A yellow dog nosed its way across to one of the shacks. After a while a heavy figure appeared on the veranda of the homestead. Malone stood there talking to his manager, Welles. He strode to the ground, tramped across the close-packed earth towards the office where Deane sat watching him come.

Malone pushed open the door. He dropped down heavily into the chair behind the desk, flung his hat on a cupboard. He sat for a while, broad and bulky, watching Deane from narrowed eyes. Then he said, 'So you're still around.'

'Looks like it.'

'I'm told that you left. And you came back.'

'Couldn't make a break from the charming place.'

Malone kept staring at Deane, twisting a button on his coat with thick fingers.

Deane drew up his legs, waiting. 'Well . . . What's the idea of asking me over?'

Malone didn't reply. He was a man of rough and violent action, in a district where there wasn't much need for talking; he seemed to be planning his words. He was quieter than he had been on their last meeting in the pub. But the red glint in the corners of his eyes was still there.

Deane said again, 'Well . . . ? I haven't got all day.'

Malone's face tightened. Then he released it. 'In a hurry?'

'I'm supposed to be out of this district by sundown. Your friend Lawrence barred me an hour ago.'

Malone's black eyebrows lifted. He said slowly, 'You going?'

Deane grinned. 'Well, I don't think so. Not just yet, Malone.'

'You're a bleeding fool, aren't you?' Malone was still twiddling the button on his coat.

'I am.' Deane said softly, 'You're right, I'm that.'

'Just what are you after?'

'I'll tell you what I told Lawrence. That I want to see you on trial for the murder of that abo. Thomas Clancy.'

Malone shook his big head, almost bewilderedly. 'Are you crazy? Killing an abo isn't murder. By God, you don't know them!' He struggled to be reasonable. 'Listen, Deane, at one time in this district a man could get himself a licence for shootin' abos when he wanted.

They had the right idea once—the blokes who cleared this country for us. They knew they had to treat abos like the dingos, like all the other animals which ran wild here before them. Clear them from the waterholes and the plains . . . They had to drive them out and finish them, so the whites could make a land out of this flamin' wilderness. They did the job well—and by heck, I'm not letting it all go back again.' He pulled himself up to his feet, heavy hands on the desk. 'Abos aren't like white men, Deane. You're a stranger round here or you'd know. White men's rules don't apply. If you ran a station, you'd find out soon enough.'

He was still, watching the effect of his words. He had spoken frankly; Deane stared back at him and knew he believed it that way. Deane moved wearily, driven and yet bitter and impatient against the unknown thing which drove him; bitter because he saw only futility in behaving the way he did. At last he said unemotionally, 'I didn't make the rules, Malone. Your own government made them. To kill an aborigine is still murder . . .'

'City men. They don't know the conditions.'

'But I haven't met anyone else with your opinions, anywhere back down the road. The abos are a dying race.'

'Are they hell . . . The others'll find out when it's too late.'

Deane said, 'You're wrong, Malone. So it doesn't alter things.'

'What things?' He was fierce.

'I won't leave this district until I see you sent down on a charge.'

Malone's eyes swelled. 'By God,' he said. 'You're

asking for it. You come into my land, you—' Then he lowered himself back into his seat. 'All right. I didn't fetch you here to argue with you. I didn't expect you to see it my way. I brought you here to talk the only kind of language you'll understand.'

'What language is that?'

Malone's eyes passed over him, contemptuously taking in every detail, face and clothing and split shoes. 'Money . . .' He spat it out.

'Go on,' Deane said slowly.

Malone said, 'You can't do anything to me. You're only a crapping little louse—I could swat you with one hand. But I'm going to get rid of you the easiest way. I could finish you off and it wouldn't trouble me a wink at nights. And nobody would know. But instead I'm going to settle it quietly—' He broke off.

Deane began to pull himself up in the chair, and the cold bitterness moved over him again. He said dully, 'How much, Malone?'

Malone said, 'I'll give you a hundred and fifty quid to climb on your bloody truck and get to hell out of it. Right away, pommy, with no looking back . . .' His voice rose.

'A hundred and fifty.'

'And think yourself lucky.' His words were thick and distorted.

Deane sat still.

Malone unlocked a drawer in his desk. He fetched out a wad of notes, began to thumb through them, licking his finger to count. Abruptly something in Deane's stillness struck through to him. He looked up. 'Well?'

Deane got to his feet, dirty and unshaven, derelict as ever. A fresh wave of bitterness passed over him. He said, 'Keep your sodding money, Malone.'

Malone's eyes opened slightly. Deane noted his face, adding it to the store of all the other looks on people's faces when they gazed on him: the opinion they all had of him.

Malone said, 'What—'

Deane crossed by him to the door. 'You won't buy me off. I told you what I'm going to do. I'm sticking to it.'

'Then you're a dead man.' Violence burst out.

'Not yet,' Deane said. He walked through into the late sunlight, standing rigidly in the yard.

Malone crashed through the door after him. 'Now you'll find out. You'll see, pommy.'

Deane said, 'I'm going back to Galah. Get your bloke to take me in the car, will you?'

Malone choked, taken aback. 'You've got a crappin' nerve. You walk. You just walk it. And if you pass out on the trip, maybe that's the best way out all round . . .'

The light in the sky was turning red towards sundown, reflecting over the ground. Deane turned away, striding forward recklessly. 'All right. I'll fix up some transport for myself.' He paced towards the homestead.

Malone came behind him. 'Keep off my place. Just get off my land.' He reached for Deane's shoulder. Deane brushed him away and began to run. Malone thudded after him. He reached the homestead veranda, raced up the steps. Suddenly Nancy Malone stepped out of the living-room and confronted them. She stared from one to the other, breathing quickly.

She said, 'What's this? What's going on?'

Deane stood still. Malone came up, lunging. Deane dodged him. Malone's wife pushed herself in between them.

She said, 'Stop it! Stop it, Richard!' Her voice was high and anxious. Malone moved forward again, then held himself back. He swallowed, his throat jerking.

They were silent, staring at one another in the reddening light.

Nancy Malone said again, 'What's going on?'

Deane looked at her. She was afraid, her cool eyes disturbed. She stood awkwardly, legs apart, staring from one to the other. He explained, 'Just a call upon your husband. I asked him for a lift back to Galah afterwards —seems to have upset him.'

Malone said tensely, 'Get going. Start walking, right away.'

'I'm waiting for a car.'

Malone moved forward again. His wife grabbed his arm. 'Stop it. Give him a car. Let him get back.'

Malone turned on her. 'After what he's said to me. After what he's trying to do . . .' The fury swept his face. 'I'll see him—'

'Give him a car and let's finish it.'

'Just keep out of this.' He pushed at her.

Then suddenly they both stood still, staring at one another. Deane watched them closely. Malone looked away in the end. He muttered something, turned and walked across the yard. He glanced over his shoulder

127

once at Deane and his wife, then he went round the corner of the sheds and disappeared from sight.

Deane breathed out. He looked at the woman. 'I'm sorry.'

'So you should be.'

'Hope I haven't made things difficult . . .'

'They were difficult already.' Her voice was flat.

They stood on the edge of the veranda in the red light. Through the curtains Deane saw the china cabinet, the mahogany furniture. His eyes came back to her face, and the smooth white skin was puckered in trouble.

He turned aside. 'Suppose I'd better—'

She interrupted. 'Take a car.'

He hesitated.

'You know you can't get back without one. It won't alter anything with him anyway.'

Deane was still hesitating. She said restlessly, 'Oh, you fool. Why don't you get to hell out of here?'

He was silent.

She walked past him, across the yard. 'I'll get my Land-Rover . . .' She drove back, pulled up beside him. 'Take it.'

He shook his head.

'Take it, damn you.'

'How will you get it back?'

'I'll send someone over to pick it up in the morning.'

They gazed at each other, she from the driving-seat, he looking up. At last he said, 'What about you?'

'Well, what about me?' She was sharp.

Deane said roughly, 'I want to know how you stand with him, Nancy.'

Her eyes pulled themselves away. She was uncertain. Then she said, 'Get in.' He glanced round the yard. She said again, 'Get in.'

Deane jumped into the seat beside her. She let in the clutch and drove away from the homestead, across the track and on the Galah Road as the dark came down.

10

SHE drove fast, silently, the beam of the lights ripping through the night. The Land-Rover bucketed across the road, crashing and bouncing, racing across the paddock. At the gates Deane jumped out, flung them open.

The truck passed through. A couple of miles farther on he said, 'Now stop.'

She swung the vehicle off the road on to a hard patch of ground, switched off engine and lights. They sat still in the dark. She said sharply, 'Well?'

Deane paused. Then he said, 'Got a cigarette, Nancy?'

She reached forward and took a pack from the compartment. The lighter flared; two red spots gleamed in the dark. He saw faintly the outline of her face. The bush was black and silent, stirred by a slow, faint wind which was cool and sour. On the horizon there was still a red tinge, a falling glow.

'Well?' she said again, resentfully.

Deane exhaled, sighing. 'Rough on you, Nancy.'

'What do you want to talk about?'

He said, 'Fierce against me?'

'Yes. I am.' He could not watch her face. Something in her voice was unsteady.

'I don't blame you. Rotten spanner in the works, aren't I . . . ?' He was silent. Suddenly he said, 'You know why I walked out on you yesterday?'

'*I* walked out on you.'

'We won't argue. You know why I wanted to break it up?'

'I'm not greatly interested.'

Deane smiled. Then he said, 'I found myself getting soft about you, Nancy. That's no good for someone like me.'

She didn't answer for a while. 'I don't suppose it is.' Her voice burned. 'You'd had your fun, hadn't you . . .'

'It wasn't like that.'

'We won't start sentimentalizing.'

Deane stared at the dim outline of her face again. 'Far from it.' The vast stillness of the bush pressed around them, the great night.

She said, 'Nothing to get sentimental about anyway. Pretty brutish, wasn't it?'

Deane looked away. 'I do some pretty brutish things . . .'

She turned abruptly. Her perfume passed over him. 'We're wasting time.' Her voice was deep and restless. 'What do you want?'

Deane said, 'All right, Nancy . . .' He hesitated. 'Don't mind me asking this. I want to know how relations stand between you and Malone.'

'Personal, aren't you?' She drew on the cigarette, exhaled quickly.

'I know it's no business of mine—'

'Too right,' she snapped.

Deane said tautly, 'I'm asking for your own sake, Nancy.'

'Mine.' She turned round. 'How in hell are you concerned?'

He caught the anger in her eyes, burning under the starlight. He said grimly, 'Because I'm taking Malone on. They may think I'm a useless blasted bum, but I'm taking him on. And it's him or me.'

'What's the matter with you?' Her voice was unsteady again. 'What's all the damn' fuss about? You'll only finish yourself, and the rest of us with you.' Her fingers played with the knobs on the panel, twisting. Deane remembered the way Malone's thick fingers had played restlessly with the coat-button.

He stared away, back to the ivory shape of her face. 'Tell me, Nancy. Just let's get it straight.'

She looked back at him. 'You're strange. The more I see of you, the stranger you are . . . Even your voice is different . . .'

'Just tell me what he means to you.'

She relaxed and sat back in the bucket seat. She dropped her head. The starlight caught the bended shape of her white, plump neck and shoulders, the short hair . . .

Deane sat still.

She said at last, 'Well, we live in the same place. Use the same rooms. But that's just about all—so now you know.' She stared at the black dials on the truck panel, fixedly. Her voice deepened again. 'He knows what I think of him. He doesn't come within a yard of me.'

'Why?'

She said rudely, 'That's enough for you. Keep your nose out.'

'Just tell me.'

She shuddered, releasing. 'He's filthy. Those dirty black beasts . . . Coming to me after them . . .' Her shoulders shook and she trembled.

Deane stubbed out his cigarette. Then he laid a careful arm about her shoulder. 'Sorry . . .'

She jerked her head upright. 'I'm going to leave him.'

'If you feel like that about his habits, why haven't you done so before?' He was hard.

She said, 'I don't know. Oh, I don't know.'

Deane tightened his fingers on her shoulder. 'You're a fine woman, Nancy. What's your trouble?'

She shook her head. He caught a glimpse of her cool, set and unhappy face in the dark and the starlight. At last she said, 'This district, it finishes me. I'll go back to Queensland one day. Where I come from there are cornfields, rivers and grass . . .' He saw her run her hands over her body, white, live and straining things in the night, over her breasts, stomach and strong flanks, her thighs. She said, 'I'm still young, I've a good body . . . This dry waterless land is no place for me. Men of bones and skin, lusting after greasy black gins . . .' Her voice was almost hysterical.

'There, there,' Deane said. Then he smiled. 'Well, I'm not so skinny or old. And I always stick to white women . . .' He stroked her shoulder.

She turned her body in the seat and faced him. She looked at him, and her face was the way he had seen it before, passive, remote and yet intimate, open to change. Cold and still, and yet ready to burn, violently and desperately to burn.

Deane began to bend towards her, his hand sliding down her spine, and then suddenly he thrust back. He took his hand from her shoulder. He said gently, 'You'll be all right, Nancy . . . Now you'd better drop me off.'

She pulled herself up, the smooth body against the leathercloth. She started up the engine, drove the Land-Rover back on the road to Galah.

The night rode by. She stopped on the edge of the township. Deane got out and stood at the side of the track. 'Thanks, Nancy,' he said.

She sat staring back at him silently. Then she reversed the truck, drove back towards Clancy Rock.

Deane turned off and walked the remainder of the way down the street. He walked slowly along the middle of the road, thinking.

And then his mind moved to other things, and he remembered it was after sundown, and he was still in Galah . . . The street was empty and the few lamps shone.

He didn't think the policeman would act anyway; he wasn't worried. He tramped up the steps to the hotel, got a couple of cans of beer to take the dryness out of his throat and body.

During the night he woke abruptly. He began to throw himself to one side, to get his feet on the floor; but he was too late. A group of men stood round his bed, striking out at him with bare fists and sticks.

Black, unidentifiable shapes in the dark; Deane struggled under the net, trying to defend himself. There was a blow on his head, on his cheek, a swipe at his arm. Dazedly he tried to force a way up, to save himself. He got a foot on the cold floor: then he was punched in the stomach. He grappled with one man, reaching for his throat, grunting. Someone else hit him across the kidneys from behind and he groaned.

He saw about five of them, and they beat at him silently, panting with the effort. Blood was running down his face. Just the dark and the blurred shapes, the panting and the thud of blows on his body ... He heaved, desperately began to force himself across the room. Then there was a fiercer crash on the back of his head, and his eyes saw brilliance in the dark, and he fell down.

When he awoke the room was light, and the time was morning. He was sick and muzzy, and he lay quiet. He could feel crusted blood on his face, from forehead, nose and lips, split cheeks over his cheekbones, bruises and a swollen eye. He could only dimly see. His neck hurt and he was full of nausea.

At last he got to his feet groggily, pulled on his trousers and staggered down to the washroom. There was no water. He went to the dining-room, pushed inside, almost toppling.

The waitress was setting the table. Through his clear eye he watched her face change at the sight of him. 'Oh,' she said. 'Oh—' She stepped towards him, hands held out, small face aghast.

Deane said, 'It's all right. It's all right, Elsa. Just

134

came to see if you could spare a drop of water—' He hung on to the door-frame.

She grabbed his arm, began to lead him to the kitchen. She gabbled something in Czech. 'What has happened? You're injured? Come and let me see . . .'

'Okay, okay.' Deane let her take him. 'Been in a fight, that's all. It'll wash off . . .' He mumbled, lights swirling about in front of his eyes.

They went into the kitchen, to the old sink. The cook was there, a red-faced, disagreeable-looking woman. 'What's going on—?' Then she saw Deane, and stopped. She stood still and watched silently as the waitress poured out a small bowl of water and cleaned up his face.

He began to feel better. He stared in the cracked mirror as the contused skin appeared under the coating of blood. He pushed back his hair, held the wet rag to his aching forehead. He leaned against the wall for a few minutes. Then he was almost all right. A black eye, a collection of bruises and gashes. He'd recover . . . Now he could think about it, and about the men who had done it. Town men, or those from Clancy Rock? His swollen mouth hardened.

The girl said, 'What was it? What was it?' She was quick and worried, warm and sympathetic. Her hands rinsed the rag in bloody water, offered it again, tended him . . .

Deane said, 'You'd better get back to your breakfasts, Elsa. I'm all right now.'

'Tell me what happened.'

'Somebody doesn't like me.'

'When did they do this?'

'During the night.'

'Not in the hotel?'

'Yes,' he said. Her mouth opened in horror and fresh anxiety. He smiled crookedly. 'Forget it.' He stepped away from the sink.

The cook said, 'If you're through we'll get on.' Her voice was flat, neither for nor against him. She forked the steak on the grill.

Deane said, 'I'm fine. I could even eat some breakfast.' He went outside, walking firmly, sat at a table.

Elsa brought him the steak and eggs. Nobody else was in the room and the kitchen door was closed. She hesitated, standing beside the table. 'You're sure you're all right?'

'I'm sure,' Deane said. He glanced up, and smiled painfully. 'You know?'

'What?'

He said gratefully, 'You seem to be the only person who puts up with me in this town.'

Her face was still dubious, but she smiled back. 'Don't be silly.'

'I tell you you're the only pleasant character I've met. And you a Czech . . .'

She shook her head. 'Now I'm a New Australian.'

'That's their gain.'

Again she shook her dark hair. 'You are not fair to the Aussies—they are good people. Fine people.'

'I know. Some of them beat me up last night.' Deane grinned briefly. Then he said, 'Good Aussies? Hell, I know that already, Elsa. I'm the sort with a grudge against everything, but I've no special hard feelings

136

against the rest of Aussie-land. I've met plenty of the best. But in other towns, not in this one. Not in this little corner they call Galah . . .' He curled his lip. He drank coffee, and vigour came back.

The girl said simply, 'I like the town. The quiet, and the bush outside.'

'Not such a bad town.' He repeated, 'It's the people.'

'Under different circumstances they would be different.'

Deane stroked his inflamed eye. 'I've something to repay now.' He began eating the steak; maybe it would give him strength, red meat . . .

When he went out afterwards he met the policeman coming into the hotel. Lawrence's face darkened, seeing him. He opened his mouth to attack Deane, then it stayed open. He peered at Deane's split face.

He stared back. 'What in creation—?'

'Don't tell me you weren't one of them.' Deane jeered.

Lawrence said, 'What's happened?'

'You know.'

'I flamin' don't know,' Lawrence said. 'What's it all about?'

'I got my sleep interrupted last night. Four or five jokers who didn't like me.'

Lawrence stared again. 'Who were they?'

Deane shrugged, pushed by.

Lawrence said, 'I didn't have any knowledge of it.'

'All Galah must have done. There was noise enough . . .'

'I didn't know.' His voice came quickly. Deane looked at him. Lawrence's expression was open and angry. But for once it showed something more than hostility towards Deane. Something in the eyes between the creased lids was ashamed; they were the eyes of a man made uneasy, a man of a certain honour disturbed by alliance with things he didn't like. He paused. For a moment even Deane thought the copper was on the edge of an apology. Then Lawrence tightened himself up. 'Going to take their hint now and get out?'

Deane shook his head.

The policeman's mouth screwed together. 'I'm just choosing my time to jail you. You'd be safer. You'd still do best to push while you've chance.'

'You've sung the same song too often, you windy bluffer. I'll go when I'm ready.' Deane said tautly, 'Arrested Malone yet?'

Lawrence's hard, lined face watched him, trembling on the brink of impotent fury. He swung round. 'No, I haven't.' He swallowed thickly, pointed. 'See over there . . . ?'

Deane gazed across the street, between the gaps in the shacks towards the horizon, the far, low hills in the morning. Straining his clear eye, he saw faintly puffs of white smoke rising, away on those distant hills. 'Smoke signals?'

Lawrence said violently, 'Yes. The local tribe is back from walkabout. They're due to come in soon for their rations.'

'Well?'

Lawrence's eyes lanced against him. 'So I know the abos are harmless enough, sport. I know I can handle them . . . But they've got some outlandish customs about

138

them still. The spearing days aren't all that long back. We stick together here—we have to. And you're a destructive influence. You're dangerous . . . I reckon that's a good enough reason to hold you. I reckon the magistrate will have you put away. Especially when I tell you the magistrate is Richard Malone . . .' His voice was fierce.

'Come and catch me.' Deane walked on down the street, ignoring him.

A noise came from the sky. He looked up and saw a plane go by glinting in the sun, a lone high-wing aircraft passing across the blue heaven, dropping lower: heading towards the airstrip at Clancy Rock Station.

The weekly mail plane, coming in from the outside world.

I I

HE sat on the edge of the rail and kicked his broken shoes in the dust, swearing roughly to himself.

Deane was heavy with a vast mood of depression. His face throbbed and he was sick. He was tired of talk, tired of constant bouts of sparring with the policeman, tired of his uselessness. There was nothing he could do. He had accused Malone point-blank, and he had shot his bolt. If the copper wouldn't act, nobody else would . . . Once again his temper and spirit had swung low, and he was sunk in futility. The initiative was gone from him and it was left to Malone or the copper to attack . . .

Again he wished he had gone, fled away with his tail between his legs.

His boot-heels dragged through the sandy soil. The sun at noon: the spasmodic life of the township moving about him. Hot sun upon his face and he sat still, head low.

A car came clattering over the dusty road, stopped in the square. It was the Chev from Clancy Rock. Deane raised his head. A man got out of the car, crossed to the hotel. Deane felt a shock go through him.

He drew in his breath.

The man carried over his bag, reached the steps. He looked at Deane, came towards him. He stopped, standing over him. Then he said softly, 'Hullo, Johnny.'

Deane was silent. Then he said, 'Hullo, Kendrick.'

'Nice to see you again, amigo.'

'Must be.' His gaze flickered upon the dry, stooping figure in the city suit marked with Territory dust and creases, the sallow and watchful face and the darting black eyes. A sense of doom went through him. He hadn't run far enough.

Kendrick said, 'We must talk, no? First I'll check myself in and get a room.'

'Share mine,' Deane said.

The leathery cheeks wrinkled as Kendrick grinned. He took his bag and disappeared inside the hotel.

Deane sat without moving.

After a while Kendrick came back. 'Next bunk to yours, Johnny. And nowhere to get a wash—you picked a damn' fine place.'

'Been through my grip?'

Kendrick said, 'You're not stupid enough for that.'

Deane got wearily to his feet. 'Come inside and I'll shout you a grog.' Now he was without emotion.

'You can afford to,' Kendrick said.

He followed Deane into the bar. They stood near the passage, glasses in hands. Deane glanced over his shoulder; the hotel-keeper had gone back to the other room. His eyes passed over Kendrick again, studying his thin and sharp figure, the crafty head. He had last seen him three months back amid the five o'clock rush of traffic in George Street: wide-brim hat pulled down, his nicotined finger jabbed forward in warning, in accusation. He hadn't changed much. Deane had altered more.

Deane said, 'Come in on the mail plane?'

'Yes.'

'For a holiday?'

Kendrick grinned again. 'Had quite a useful talk over at the airstrip before I hitched a lift here, Johnny. Met up with a big guy named Malone—I hear you've stuck your neck out quite a bit, these last few days. You've got a lot to lose . . .'

Deane stared at him sharply.

Kendrick said, 'So I told friend Malone I was as interested in you as he is. We put ourselves on quite good terms.'

'Quick work.'

'Yes.'

Deane swallowed his drink. At last he said, 'How did you find me, Mex?'

'The boss sent me after you. He doesn't intend to write off twenty thousand just like that, Johnny.'

'I tell you again I never got near the bloody money...'

Kendrick ignored it. 'I've been following you all the way across. You chose a rough road, amigo. You nearly killed me...' He laughed wryly. 'Anyhow. I was a way back down the track making myself friendly with the local johns when the wire came through about you. Didn't think there'd be two nasty-looking pommies around this area—and when I got the name I knew.'

'I ought to have kept it quiet.'

'And missed this old acquaintance?'

Deane finished the beer, the last dregs. 'You've wasted your time. I ran from the police, not from you. I ran because I was a damn' coward, not because I was guilty. I've got nothing.'

'You'd better have. It's turned up nowhere else.'

Deane shrugged. 'That's not my worry. If I had twenty thousand quid I should travel a different way from this.' He wiped his mouth.

'Maybe. Maybe not.'

Deane stood at the door, staring out on the street. 'This is the last time I'll repeat it—I never touched a penny. You may as well believe it. I've too many other things on hand at the moment to care a blind curse about you.' He said shortly, 'You'd better go back where you came.'

Kendrick shook his head.

'Then do what you like. I'm not bothered.' His mind was empty.

Kendrick turned to him. 'I'm hanging on to you until the money turns up.'

'Hang on,' Deane said. He went back to the counter

and ordered another beer. He stood by the bar, gulping it down.

Kendrick eyed him narrowly, his face dark against the light outside. Then he said, 'I'm watching you all the time, amigo. Cough up and we'll part as friends. Try to get away with it, and you'll find me on your tail. And you'll find me rougher than you've known me.'

Deane banged the glass down on the bar, crossed over to him. 'I'm in a bad mood. Just leave me alone.' He walked out and down into the street. He heard Kendrick step on the veranda to watch him.

He strode rapidly down the street. Now suddenly he was in a rage to act, to fight back. He caught up with the old aborigine Albert, limping towards the wurlie camp. Deane stopped him. 'Come on. We're going for a ride, Albert. We go fetch 'im that feller Jack, tak 'im see policeman. We go now . . .' He spoke savagely.

The abo looked uncertain. 'Aw boss? That feller policeman, him no—'

Deane interrupted. 'We're going. We'll set the bloody thing boiling. Come on.' He gestured to his truck.

They crossed over the square. Mex Kendrick's thin frame draped over the hotel veranda rail, dark suit and big floppy hat, watching them.

The truck fired, began to move forward. Deane raised a mocking hand, spun the wheel and they rode out. Kendrick stood shading his eyes.

The bush was sterile under the midday sun. Heat rays glittered over the rock and mirages shone upon the plain. Still trees and heavy air, a goanna lazing on a boulder.

The distant hills, where the white puffs of message smoke drifted up, the dead mulga and then the burning outcrops where the abo lay.

The bush was empty. Deane pulled up the truck, and they climbed to the crevice. The abo lay sick, moaning and turning. Deane looked down at him. The black skin broke forth into sweat, the body swelled.

He stared at Albert. 'We catch 'im plenty quick. We tak 'im now.'

Albert nodded. They carried Jack Clancy down to the truck, placed him inside. Deane began to journey back to Galah, over the stony flats. He sat staring out from the cab of the Austin. He licked his dry lips, took a swig of water. They came back on the road, travelled into town across the high afternoon.

He drove down the street, stopped once more outside the policeman's shack. He said to Albert, 'Wait.' He got down and went to the office door. Lawrence was not there.

He went back to the truck. The wounded abo moaned mournfully under the hot canvas roof. A couple of children collected around. He looked across at the hotel and Kendrick had come to the steps, was staring out. Margie Thompson was at the door of her store. The post-office man waited. The township watched him, the way it had done when first he had driven in . . .

Flies swarmed. The sun blistered over the scene. Kendrick came down from the veranda and joined the spectators around the truck. Nobody spoke. Deane licked his parched lips again, mopped his sweaty forehead. Old Albert sat fixed and unmoving in the cab.

144

Lawrence came into view, walking round from the back of the square. He caught sight of the group around Deane's truck: he kept his face bare and walked stiffly on, up to the office.

He cast a quick glance over the spectators around the truck and the body inside. Then he turned to Deane. 'What's this?'

Deane jerked a finger at the wounded aborigine. 'Someone to see you. He's a sick feller, Lawrence. He's had a rifle bullet through his stomach. But when he's well enough he'll make a statement to you.'

Lawrence's gaze scanned the aborigine and then came back to Deane. 'What sort of statement?'

Deane said calmly, 'He saw the men who hanged Thomas Clancy. He's an eye-witness. They shot at him . . .'

Nobody spoke among the spectators. Quiet lay flat under the heavy air and the sun. Deane was conscious of the sun as a great fire in the sky, burning on his back.

Lawrence bent over the aborigine. The bony body quivered and tossed and the big-lipped mouth writhed. He looked up. 'He's delirious.'

'He wasn't when he saw it happen.' Deane said, 'Get him better, Lawrence.'

Lawrence paused. Once again he glanced round at the onlookers. Then he said, 'Carry him in.'

Deane and the old aborigine carried the wounded man into the station.

The policeman motioned to the back door. 'Through there.'

They laid him on a bunk in the barred rear room. His

eyes rolled from side to side but he did not seem to be aware of anything. Deane asked, 'Going to call in the flying doctor?'

'I'll see how he goes.' Lawrence said, 'Leave him to me.'

'He's got a statement to make. I want him well.'

'I'll look after him. I'll call in the doc if necessary.'

'Sure?' Deane said.

'I'm sure.' The policeman's face went fierce with anger. 'He's in my care now.'

Deane stared at him. Then he said, 'Very well, copper. He's in your care.'

Lawrence met his eyes. They went out to the front, standing on the veranda. The policeman waved his hands, spreading them, gesturing to the onlookers. 'All right, all right.' They dispersed. He said to Deane, 'Now take your flaming truck out of the way. Take it round the corner and hide it out of my sight. I never want to see it again.' He burst out abruptly.

''Fraid I'll be bringing in any more abo bodies?' Deane laughed. 'Nuisance, that's me . . .' Then his voice quietened. 'But this is something you can't ignore, copper. You can't overlook a witness's statement—abo or not. You'll find out.'

'If he gets fit.' Lawrence stuck a finger in his belt.

'He'd better get fit.'

'Where's he been these last few days then?'

Deane was silent.

Lawrence said, 'Maybe you've brought him to me too late, pommy . . .' He went back into his office.

Deane watched the green-shirted cattle-hand he had

146

fought with a day or two back, cross over to the Chev and drive her out of town. Malone would soon know about the abo.

Kendrick waited until the dust settled; he snickered and crossed over to Deane's side. 'Don't know what you're after, Johnny. Playing a risky game, aren't you?'

'Just keep away from me.'

Kendrick said, 'I guess the john here doesn't know you slipped the smoke in a hurry, eh? Shall I tell him the Sydney cops would still like a word with you?'

'Tell him what you like.' Then Deane turned on him. 'I've got enough on my mind, Kendrick. Attack me and I'll pull you right down as well. I'll finish you.'

'Cough up, and let's quit.'

'Get to hell out of it, you stupid bastard. Get going.'

He pushed past and walked on. Then he looked round. Kendrick had a small knife in his hand, sheltered in his palm. He crouched slightly, menacing; then he flicked the blade away into its socket, straightened up again.

Deane repeated, 'Get going.' He walked a few yards farther down the street. Kendrick was following him on quiet feet, smiling. He stopped when Deane stopped. Then he followed on again. Deane walked into the café and got a cup of tea and a sandwich. Kendrick bent his long thin figure into one of the cubicles and waited.

The café owner stared from one to the other, smoothing back his small moustache in uncertainty.

Deane finished and came out. Kendrick pulled himself up and followed. His mocking face fixed on Deane

147

and they went down the street. He said, 'I'm going to wear you down, amigo.'

'You'll find out, Mex.' Deane kept on walking. He was soon sick of the whole blasted thing and he went back to the pub to get some cans of beer, to get drunk and forget it, with Kendrick's quick black eyes upon him.

He sat on his bunk with the cans and an opener. He leaned back against the whitewashed wall and sighed. Then he began to drink. The first beers were ice-cold from the refrigerator and then the heat in the room turned the others tepid.

Kendrick squatted on the other bunk. Deane said, 'Have one?'

Kendrick stretched out a hand. 'Why not, Johnny? If they're on you.'

Deane knew his weakness from before. He watched the sharp head go back, the skinny throat swallow greedily, the bone jerking. Soon Kendrick was drunker than Deane was, rolling over to the bed and lying still.

Deane felt tight to bursting with it. He staggered outside, sat on the back porch to recover. He stared out muggy-eyed over the backyard, the litter and empty cans, broken bottles and the smelly privies; and the empty bush beyond, flat and dead.

The waitress came out from the back entrance of the hotel. Through swollen eyes he watched her smile; then a look of concern came over her face and she stepped towards him. 'You're not well again—'

He shook his head.

She bent over him. 'You look—' Then she caught his loaded breath and realized. 'Oh—'

Deane said, 'I'm lousy drunk, Elsa. Nothing else . . .' He reached out for the veranda rail, turning his head away.

She grasped his shoulder. 'You shouldn't—'

He brushed her away. 'I'm lousy drunk. Just see old Kendrick. Old Mex from Tampico. He's drunker . . .' He laughed. Then he began to feel for the veranda rail again, to draw himself round. 'Listen, Elsa—' Something important came to him suddenly.

'What?' She stood on the bare sandy earth of the yard, watching him gravely. She had an apron round her and her small hands were dirty. Her face wobbled and shifted before him but his gaze fastened on the brilliance of her violet eyes.

Deane said, 'I'm going, Elsa. I'm pulling up my stakes and goin'. I'm runnin' out, right out. This crummy stinkin' place. I'll leave them all behind . . .'

She didn't speak.

He reached out for her. 'Listen. You come with me, heh?' The unexpected idea had moved him. 'You come with me, down the road . . . You're a good girl, I like you. You're a girl of—of—qual'ty. You come with me Elsa, heh?' He struggled to focus his eyes.

She took a step back. For a moment he didn't understand her face and he blinked in the sunlight, trying to concentrate. Then she smiled calmly. 'No,' she said. 'No, Johnny. I'm not coming with you. You run out alone.'

'Come on.'

'No.'

He said suddenly, 'I love you, foreign Elsa. Nobody else . . .'

149

'You're drunk.' She smiled again, with cool cheerfulness. 'Now I am working.' She walked past him as he sat, her skirts rustling, into the hotel again.

Deane sat staring at the dry earth and then he flopped over and slept in the sun, breathing heavily.

When he got up he had a killing headache. He was roasted hot, dry and nauseated again, and the wounds on his face ached. He knuckled his hands over his forehead, groaned and went back inside.

He sat on his bunk. Kendrick was still out on the other bunk, snoring, ugly and sour as he slept. An Australian, although he had lived years in Mexico and been jailed in most other countries: drab-skinned, unhealthy, with pouches under his eyes and a cruel curve to the tight purplish lips of his small mouth.

Deane ran his hands over him. He found nothing to interest himself. Then he searched through Kendrick's bag. He discovered a .45 automatic. He checked it over, then stowed it back where it had been. Afterwards he slapped Kendrick across the face a few times. Kendrick came round.

'Drunken hooer,' he said, moaning at himself.

'Feel good?'

'Gaw.' Kendrick bent over and rested his eyes.

Deane said, 'You passed out. I could have been off and gone if I'd wanted to. But I've no reason to run. I've got nothing you want.'

'You were cut as a turkey yourself—you couldn't run. Anyway, you knew you wouldn't get far . . .'

'Stop wasting your own time.'

'I'm sticking to you.'

Deane said, 'I'm getting annoyed.' They sat on their bunks, bleary-eyed and sick, staring at one another viciously.

Later Deane went over to the store. He paced down the street; Kendrick pulled himself to the veranda and watched him go inside. He started forward, then clung dizzily to the post.

Deane walked into the store, sat down limply on the top of the counter. The ache in his head burst through him as he sat, pain built out of his face injuries, strong sunlight on swollen eyes, the heaviness of beer.

The fat woman eyed him warily. She put her hand on her hips. 'What do you want?'

'Let's have some aspirins.'

'Going soft?'

'I'm sick today, Margie. I'm bad.' He was sorry for himself.

'You ask for these things.'

'All right, all right.' He took the bottle, helped himself to some tablets, crunching them in his mouth.

He looked up at her, waiting in the shaded shelter of the shop. He caught the doubt in her eyes, the lack of warmth. He stirred restlessly.

He said, 'What's the matter, Margie?'

'Nothing's the matter.'

'You don't seem so friendly.'

'Hell,' she said. 'Friendly ... You want me to slobber over you? You want me to stroke your forehead 'cause you're crook? I heard what happened last night and I'm sorry, Johnny. But I got no sympathy for you. You're a dirty, beer-swilling no-good and you deserve all you get.'

He studied slowly her fat face with its stern expression; his gaze dropped to the heavy ankles and loose slippers. 'I only drink when things go badly, Margie.'

'That's the worst time. That's when a man should fight back.'

'Easy to talk.' Suddenly he began to straighten himself up, hurt cruelly. He brushed his hands over his face again, rousing himself. 'Too damn' easy to talk. When it comes a man just hasn't got the guts . . .'

'Some have.'

Deane got to his feet. She looked at him with the same critical indifference. He said, 'I'll show you. I'll show you . . .'

'Come on.'

He stopped. 'But you don't want me to fight back, Margie. You want me to fall down. You want me to cry off and run. Don't you now? You want me to push off and leave your lousy little town to its own rotten affairs. Don't you now?' His eyes were glittering.

She hesitated. Something in her face softened suddenly. She said, 'I like to see a man with courage.'

'Let's not talk about courage. Forget all that—' He swept an unsteady arm, knocking it against a cupboard. He felt the torn skin smarting. 'Forget it— But you want to see me lose out against Malone. You want to see him keep me quiet, if nothing else. Eh, Margie, eh?'

She said slowly, 'I don't know. Oh my word, I don't know, pommy.' Her voice was strained.

Deane stared at her, swaying on his feet. He frowned, and then he stepped forward. Eagerness swept him.

'Help me, Margie. If you stand up against Malone, maybe the rest of the town will. Just help me . . .'

'Help you?' The scorn came back. 'What's it to do with you? You, who only wants to smash us all up. You've told me, pommy—don't forget your own words . . .' The harshness was over her again.

'I know. I know . . .' He shook his head dully. 'I know it all . . .' He stared up at her again, peering through the cloudy swimming of his eyes. 'Now I'm trying. Believe me, I'm trying.'

She stood still. One hand rested uncertainly on her big hips. She moved slightly, staring back at him.

Deane said, 'Give us some help. It'll make all the difference if you're around . . .'

She said, 'You're still drunk. Go and sober yourself up.'

He turned for the door, stumbling.

Behind him she added slowly, 'You can come back again to try an' convince me when you're sober. If you want, Johnny . . .'

Deane brushed his hand over his face. 'All right. I'll come back.' He reached the sidewalk, went back down the street. He walked unsteadily. Something stirred in him, under the drunkenness, one gleam of desperate expectancy.

He stepped down the street, placing his feet carefully. A young half-caste girl got in his path, tangling up with him. Deane stepped aside and checked, swaying. 'What—?'

The raggedly-dressed kid whispered to him. 'Hey, mister. Hey, mister . . .'

'What?' He blinked down at her.

'Man, he wants to see you. Man, him waiting.'

'What man?'

'Dunno.'

Deane said, 'Where?'

She pointed with a thin arm. 'Round a' back, that house.'

Deane straightened up, began to walk to the rear of the shack, off the main street. He went directly without thinking, his brain still confused, careful only that his weaving feet should not lead him to a fall.

He walked round the corner, to the back where the rear walls faced out over a dead land of scrubby bush. Then he stopped, and caution sobered him more quickly than a bucket of water in the face. Waiting confronting him was Richard Malone, and with Malone was the black aborigine stockman. The stockman carried a thonged cattle-whip; he was standing square, gripping tight to the heavy butt.

Malone stared at Deane, and then he came forward slowly.

12

HE stopped a couple of yards away. Malone said malevolently, 'You ought to've learned from last night.' His heavy face poked forward at Deane, carved by raw emotions of hate and violence. The black stockman waited behind him.

'Give me time.' Deane swayed on his legs, pulling himself up.

'You've had that.'

He felt too ill to want trouble at this moment. 'Listen—'

Malone said, 'Ken.'

The ringer's whip flicked. Deane began to duck, but even before he moved he felt a fiery sting at his left ear. He put a hand up to it, and the tip was nicked and bleeding.

He straightened himself. 'Drop it.'

Malone jerked his head. The whip cracked again, and Deane felt the ear rip in the same place. He stepped back; his legs were weak and he almost fell.

'Pack it up,' he said unsteadily.

Malone watched him. 'You drunken sot. And you turn against me . . .' The glint of red came into his eyes.

Deane raised an arm warily to protect his face, glancing from side to side for a way out. 'Let's call this off.'

Malone said, 'And you're going to get this too for bullin' around my wife.' Throaty fury deepened his voice. 'That alone gives me the right to kill you . . .'

Deane's gaze lifted. Malone was burning with a hard rage which lay in the creases of his forehead, the pulled mouth and gleaming eyes. He held himself fixedly upright, khaki-breeched legs planted apart, a rigid set to his head on the thick neck. He stared at Deane with tyrannical, vengeful face.

Deane said, 'Where does your wife come in?' He started to retreat.

Malone stepped after him. 'You know. You know, you stinking whore-chaser.'

Deane licked his lips. The beer had left his throat dry and the headache had burst back on him, pounding. Malone gestured and the whip cracked; and the thong cut deep into Deane's cheek. He winced. He said desperately, 'Do your own dirty work. Handle it yourself.'

'By Christ, I will.' Malone swung round, grabbed for the stockwhip.

Deane took the opportunity to run; he turned, began stumbling towards the corner of the shack, the street. His knees bent and he tripped, sprawled. He lay face down on the soil, cursing. The whip cut, and he felt the shirt rip on his back. His hands clawed the earth.

The whip tore him again, slashing his skin. He yelled, began to scramble up. The crack of another lash sent him back to the ground, flung by the impact of the leather. His face pressed into the sandy earth. Malone was panting behind him, swearing frenziedly as he wielded the whip.

Deane's hands scrabbled for a grip to lift himself to his feet. The whip cracked down on him again and again. The pain was sharp and fierce, piercing his back. He yelled once more, trying to rise.

His head reeled, and he flopped down. He moaned, twisting under the lash. Hell, if he were properly sober ... The lashes came down.

Abruptly his mind changed. He lay still now under the cuts, concentrating with dazed brain. Suddenly something inside him almost laughed, bubbling up to his lips, grim and agonized amusement. The hurt of cut flesh, the beating, humiliation ... The blows fell upon him . . . Suddenly Malone was setting him free, unknowingly releasing him. A scourging, clearing his sense of guilt. He had done some lustful things since he came to Galah: now he was paying his debt to Malone.

Footsteps came round the edge of the buildings. Deane heard dimly, his face pressed hard to the earth, arms across his ears. Flat feet thumped over the soil past him. The whip came down again. Then there were voices, argument. The flailing of the whip stopped. Angry voices again, a woman's; then some other person's footsteps followed round the edge of the shack to join in.

Deane took his hands from his head. Malone's thick voice was profaning loudly. A boot crashed against his ribs and he groaned, collapsing again. Heavy feet thudded away. All was silent.

Slowly he rolled over. He lifted his arm from his eyes, carefully uncovered his gaze.

The sun struck him. He moaned again as he felt the pain in his back. Malone and the abo stockman had gone. Margie Thompson stood over him. Her big, worried face cleared against the sky as his eyes focused. She lowered herself to her knees. 'All right, Johnny? All right?'

He stared around, over the bush, then twisted his head. Mex Kendrick leaned against the shack wall, long and skinny and dark-suited, glancing down on him sardonically.

Deane began to struggle up. The fat woman said, 'Take it quietly. Your back's not so good.'

'It'll mend.' He closed his eyes for an instant, then drew himself together. Waves of nausea swept him.

157

Kendrick said dryly, 'Better thank the lady. She surely saved your chips.' His jaundiced, mournful face hung dizzily in front of Deane.

The fat woman was distressed. 'I heard the noise of the whip . . . He's gone too far . . .'

Deane said slowly, 'I asked for it. Like I always do, Margie. Now it's even . . . But thanks.'

'Watch what you're doing in future.' She was worried again. 'You better come back and let me clean you up. You're a real mess.'

Deane got to his feet. He swayed. The blood coursed through his back and he felt the bleeding flow. He bit on his lip.

She put her arm round him. 'Come on, Johnny.'

He looked stupidly at Kendrick; then he limped across to the store, sat down heavily. The woman washed his back. 'That's better. That'll heal . . .' She clucked her tongue.

'Just cover it over for me—stick a bandage on. I'm grateful.'

She wound the white stuff round his chest. 'He's gone too far. He's doing this place no perishin' good . . .' She spoke to herself.

'Hell—I wouldn't be vain enough to call a lashing at me going farther than hanging an abo.' Deane's head was swimming again. He muttered over the pain, 'But he's going to fall. He'll come to an end . . .'

She glanced at him. Then she said slowly, 'Who's your friend?'

'Kendrick?' Deane said, 'No friend of mine. He's come all the way from Sydney to make trouble for me . . .'

She bent over the pan of precious water in the small back kitchen, wringing out the cloth. He remembered how the girl Elsa had washed his face, only the morning.

'This is my bad day.' He said, 'But at least I've found sympathy. That's unexpected.'

'Had enough, Johnny?'

He shook his head fiercely. 'Nowhere nearly. I'm staying on.' Expectancy began to move him, tensely.

'You're a funny one.'

'I know that.' Deane dried his face, handed back the towel.

She said, 'Finished it ... You better get yourself a new shirt.'

'Sell me what you've got, Margie.' Deane stood up. 'Right. And thanks again.'

She said, 'Glad to do it.'

He tightened his belt, went to the hotel. His back was raw and throbbing, but he was indifferent to the pain. In the bedroom he pulled on the fresh shirt, tucked it in and walked out on the street. He walked tautly in the loud cowboy shirt, with a mixture of a swagger and yet humility about him. Boasting, because he could take punishment and not give in. And humility; because he was ashamed of what he was, and self-contempt was suddenly a hard and demanding instead of an easy thing to bear.

The townsfolk watched him past, striding down the street. They stared from behind windows, from across the road. They looked at him and their own expressions too were mixed. In their eyes was speculation, suspicion,

159

and yet a reluctant appreciation. The children peeped after him, dodging round the shacks. Old Albert nodded at him gravely, and Deane went by.

He met the hotel-keeper's wife, coming out of the store. Her quick, intent gaze swept him uncertainly and she half-stopped, hesitating.

Deane said carefully, ''Evening, Mrs. Fisher.' And he passed by.

He came back to the hotel as the bar opened at five. He went inside for just one beer, the hair of the dog . . . He drank it slowly.

The hotel-owner lingered in the bar, elbows on the counter. He eyed Deane. At last he asked, 'How're you feeling?'

'Fine.' Deane turned to him.

Fisher nodded. He studied the bruises on Deane's cheeks, his torn ear with the coagulated blood. 'Rough treatment, mate.'

There was no ill-feeling now on his lumpy, ugly face, and his tone was almost apologetic. Deane stirred. 'That's all right.' Again he felt ashamed of himself.

He met Kendrick at the doorway. Kendrick said, 'Where're you going?'

'Walking.'

'I'll come with you.'

Deane squatted on the veranda. 'Then I'll change my mind.'

Kendrick grinned. He spread himself over the rail. The dusty street lay below in the evening sun. He said over his shoulder, 'Aching, heh?'

'Plenty.'

'You've got more to come.'

'Who from?'

'Malone.'

'We'll see.'

Kendrick turned his head. 'Got anything to tell me yet?'

'You know I haven't.'

'You're unpopular enough around here—why not pull out and we'll go.'

'Where?'

'Where there's a bit of life and civilization.'

'Sick of this place already?'

Kendrick wiped his forehead. 'Hotter than bloody Mexico. Yes, I don't love it around here. Nor do you.' His black eyes watched Deane. 'You've got nothing to stop here for.'

'Unless I've twenty thousand quid buried away in the bush.'

'Supposing we split. Then I go back and say I couldn't find you.'

Deane shifted impatiently. His back hurt worse. 'You know too damn' well I haven't got the money. Don't you, Mex?'

Kendrick's narrow, jutting face was clever and cruel, and now it was amused. He looked Deane up and down. 'I don't think it's so very likely you've got it.' He smiled.

'Then leave me alone.'

'I don't want to make a mistake. I've come fifteen hundred miles . . .'

Heat began to rise in Deane. 'I'll convince you . . .'

161

Kendrick was still smiling. 'Come on.'

Deane sat back. 'Why should I waste my time? You'll get tired and go.'

'Maybe I shall.' Kendrick lit a cigarette and smoked.

The street beyond was quiet. A truck pulled up and a couple of men came into the bar for drinks, wiping their mouths with the backs of their hands. They were long-legged, dressed in shirts and dungarees, with plain, rough faces. An aborigine walked by, bare-footed in the dust, his boots slung around his neck. The sun was sinking lower.

Kendrick said, 'Where were you making for?'

'Nowhere.' Deane said, 'Just running. Just keeping moving, because I never wanted to stop and think.' His voice was bitter.

Kendrick drew in luxuriously on the cigarette, flung his head back to exhale. 'Never found your place, eh? I know, Johnny, I know. Fouled-up half my own life chasing around, because I never knew what I wanted.'

'I don't belong to your bloody type.'

Kendrick laughed. 'Your loss, Johnny. I learned a bit in every port of call. I play it crooked and rough now, and I do all right that way. Not you ... You were never one of the boys, were you? You're not an honest fool and you're not a twister, are you, friend? You're lost in the deeps between ...'

'Perhaps.' Deane watched him soberly. Kendrick dabbed the sweat and dust off his face again with a stained handkerchief. Deane said maliciously, 'And you're not doing so well after all. Otherwise the boys wouldn't have picked you for this blasted wasted wild-goose chase halfway across the country. You're not so highly placed, Mex . . .'

Kendrick shrugged bony shoulders under the jacket. 'There are other mobs. Like you, maybe I'll move on . . .'

'Just move out. Just get to hell out of it, and leave me alone. You've nothing to win.'

Kendrick stared at him closely. 'What's it worth—?'

'I own a worked-out truck and twenty quid in cash. I need both.'

Kendrick laughed. 'No, you don't convince. You're too damn' flat and finished to make me believe you, Johnny. You're not that low. You've got it tucked away, somewhere.'

Deane said, 'Ever looked at me?'

Kendrick's eyes raked him. 'I get the idea.' He laughed again. 'I don't know. I just don't know, Johnny . . . Sometimes I wonder how you got mixed up with this thing at all.'

'I wonder the same . . . But I'd still have ended up here.' His voice was barren. 'Here I am. And here I stay.'

'Here?'

'In the middle of nowhere.'

Kendrick stared at him carefully again. He threw away the cigarette stub, stepped down to the street and walked away.

Deane remained on the veranda, watching the sky change. The sky looked the same all over the world. He lifted his eyes, and the bush and the shacks and the dry land fell away. Just a blue sky, turning red. Like a blue evening sky over London, when he was a kid staring from a back bedroom window, and the world below was heady with promise . . .

He stood up. He thirsted for a few more drinks. But he couldn't afford them, and he wasn't going to end it like that anyway. His lacerated back itched and burned under the stifling bandage. He screwed up his face, walked down the street.

A small boy ran towards him, clutching at his trousers. 'Boss . . . boss . . .'

Deane stepped aside, brushing away from the piccaninny. 'What?' He was still caught up with his own thoughts.

The aborigine child in the long shirt and bare legs gabbled at him. 'Longa this way. You come this way, boss.'

Deane tightened up sharply. 'No thanks.' He brushed by.

'Yes, boss. Boss . . .'

'What do you want?'

'Boss lady. She waitin' . . .' The aborigine boy hopped on one leg, eager.

Deane stared at him. Then he glanced around the deserted walls of the town. At last he said suspiciously, 'Show me.'

The small boy led behind the shacks, to the rear of the courthouse. Deane followed slowly, looking around and prepared for fight or flight.

Near the back of the courthouse a Land-Rover was parked. Deane stood still, doubly cautious. He stared again at the young aborigine. Then Malone's wife appeared in the open doorway of the old building.

164

He relaxed slightly, nodding to the boy. Slowly he went forward. Again he looked around, at the bare silent rear walls, the deserted path.

Nancy Malone stepped inside, and he walked into the courthouse. She went down the passage into the main room with the bench and seats where they had met before. Then she turned to him.

Deane said, 'So it's you.'

Her eyes wandered over him. She was pale and still in the quiet light, her body poised anxiously. 'I wanted to see you.'

He glanced round the silent, shadowy room, and then he nodded.

She said, 'You're all right?'

'Yes.'

'I was afraid he'd kill you.'

'Nothing like it . . .' He said, 'Someone might have noticed you, Nancy.'

'I don't care.'

He paused. 'Sorry about everything.'

'I came here to apologize to you.'

He turned, startled. 'What for?'

'Letting him find out—'

'Don't be crazy.' Deane said sharply, 'The blame was mine. I got what I deserved.'

She was on edge, trembling. She came up to him, gripping his shirt. 'He'll get you another time. He'll murder you, Johnny. I know him. He's unstable . . .'

Deane took her hands, lifted them free. 'All right, Nancy. I can look after myself.'

She said agitatedly, 'He's fixing something else. He

165

and that horrible black stockman—the only abo he trusts —they were together—'

Deane said, 'Forget it, Nancy. I'll watch out.' She was silent. He stared at her carefully. She was gripped in a state of acute anxiety, shaking all over. She was good-looking and fine, but something had cracked in her. He added, 'You want to think of yourself, Nancy. You need a rest—'

'Rest?' She burst out, 'My life's all rest. Silence, and empty rooms, and black faces and the heat and drought and the silence outside. God, it's not rest I want . . .' Her hands shook and she clung them to him again.

Deane put an arm round her and she pressed violently against him. He remembered how they had swept together in frantic lust, and the variability of relationships shocked him. He felt cold as ice to her, but gentle. He was appalled and upset. He said again, 'It'll be all right.'

She lifted her face and stared at him. Her eyes were wide and her full mouth apart. Moisture beaded her high white forehead. She whispered, 'I need you, Johnny. I need you now. I need to be alive. Oh God, I need love . . .' She wrenched her head round, pointed at the long bench where the dust lay. Her eyes came hungry back to him.

Deane stared at her as a stranger. He swallowed, full of reluctant pity and shame. He started to speak. 'I—' He stopped.

'Come on.'

At last he said, 'My back's bad—' He stopped again. He watched her face change. Bitterness swept over it, fixing it in a mask of despair. The eyes beneath went

166

chill. She wrenched herself away. She stood stiff, still, trembling, brushing back her hair. She covered her face in her hands and then she began moving to the door.

He caught her arm. 'Just wait.'

She pulled away. 'Don't say anything. Just leave it. It's clear enough . . .'

'You know what I am.' He held her. 'I never pretended—'

'I know you didn't. I know just what it was. And I offered just that—' She choked.

'We came closer. The other time I was with you—'

'Oh Christ.' She shouted, 'Just leave me.'

He gripped her hard. Then he knew there was nothing to say. He let go. She pulled herself quickly away and rushed out. Her tall, firm-bodied figure passed out of the doorway. Red light from the air beyond flickered over her hair. Then she was gone. The engine of the Land-Rover revved madly, then the noise began to fade, dying down the road.

Deane stood silent in the middle of the dusty room. Rusty sunlight suffused the interior, the old books, and the footmarked sandy floor. He put out his hand and stroked the dark wood. He felt the dust between his fingers.

Slowly he walked out, went back to the hotel under the deepening sky.

13

KENDRICK snored all that night in the bunk beside him. He slept on his back, his big nose shining in the starlight through the half-shuttered windows.

Deane moved his bed to one of the other bunks. He lay awake most of the hours, expectant under the net, waiting for someone to come. But this night they left him alone. The township was dead and silent, the street empty beyond. In the morning he was tired and heavy-eyed.

He dressed while Kendrick was still asleep. Deane watched him with cold preoccupation. Then he went outside, into the fresher air of the morning. As he stepped forward his back pulled tight; the skin was scabbing, painfully tearing. The bruises on his face were stiff and sore. He felt terrible. Outside, the air was already dry, lying heavy in the early heat. The barren land stretched out to the horizon.

On the far ground behind the police station, a new encampment had crept up overnight: the bush tribe had come in for their ration issue. Smoke rose from their fires, curling to the sky. Primitive dwellings, shelters built over the fires, bark coverings; naked black figures moving through the camp-site, ebony skins shining. Deane gazed across, straining his eyes. The bush aborigines were wilder and fiercer-looking than the local crowd. They were leaner and hungrier, elemental without the lazy degeneracy of the township aborigines who had lost their own culture and been allowed nothing but the trousers of the white man's . . . The naked bodies flitted through the encampment, and away on the hills white smoke signals were still arising.

Deane watched the aborigines, and suddenly his mind saw sharply again the picture he kept recalling: the wide bush, and the glowing emptiness across that afternoon, and the desolate eyes of Thomas Clancy before he was taken away to be murdered. The terrified face of the other aborigine, before the rifle cracked and the bullet pierced his stomach.

And the faces of the other men in the Land-Rover: Malone with hard, red eyes. The young jackaroo, Gair. The black stockman with the whip. Three guilty men, and they'd all come to judgment . . .

He found the policeman inside, fiddling with the aerial of the transceiver. Lawrence plugged in, set the frequency range, then turned ill-humouredly to Deane. 'You again, blast you . . .'

'Still alive,' Deane said.

'Getting yourself in more trouble.'

'Too much.'

'I warned you.' The copper's voice held a note of unrest.

'Still alive,' Deane said again.

Lawrence stared away. He nodded at the transceiver. 'It's working now. I got some new tubes in by the plane yesterday.'

'So?'

Lawrence said slowly, 'May be handy, pommy.'

'Called in the doctor?'

'No need so far. I've notified it. That abo's coming on well.'

'He'll recover?'

'He'll do all right.'

Deane said, 'Had a talk with him yet?'

'I'll wait till he's fit.' Lawrence's voice was sharp.

'That better be soon.'

Lawrence ignored it.

Deane said, 'Can I see him?'

He hesitated. 'Take a look.'

Deane opened the door and went into the rear room where Jack Clancy lay. The aborigine was flat on his back, breathing evenly, his eyelids shut. The swellings on his face had gone down and he looked healthier. Deane went back into the office. 'Give him an injection?'

Lawrence nodded.

'When will he come round?'

'Any time.'

'You'd better watch him.'

'*I'm* watching him,' Lawrence said sharply.

'And keep Malone away from him.'

The anger came back to Lawrence's face. 'You're too damn' talkative. I've been letting it go—better not presume.'

Deane said suddenly, 'I'm serious about it, copper. I don't want to see anything happen.'

Lawrence glanced at him, probing his face. Then he said, 'All right.'

Deane crossed to the door. 'The abos are here in force.'

'Time for the handout.'

'Fearsome-looking characters.'

Lawrence said, 'They still carry their spears, pommy...' He paused, about to say something more. Then he kept silent.

Deane went outside to get his breakfast.

He sat at the table in the hotel dining-room. Elsa brought the meal. She studied his battered face; then she said, 'Good morning.'

'You're formal today.' He took the plate.

'Why?' she said.

'"Good morning." Just say, "Hullo, Johnny."'

'I will remember,' she said. She was quiet, setting the plates, the sauce bottle, looking down at the table.

Deane said, 'Eggs and steak. It grows familiar . . .'

'We have nothing else much to give you.' Her voice was low. She stood very sedately, her face delicate and contemplative. Her eyes were averted from him, her head slightly turned away so he could see only the dark, straight hair and the outline of her slanting cheekbone, a tip of nose.

He asked, 'Anything wrong, Elsa?'

She shook her head. She straightened up and for a moment she faced him. 'Of course not. Why should there be?'

'You're not smiling as much as usual.'

A brief smile touched her then. She said, 'It's early in the morning. Leave me time . . .'

'That's better.'

She went away, brought back tea. 'It worried me to hear what happened to you yesterday.'

'Then you worry too much.'

She nodded. 'I worry about a lot of things. I can't help that.'

'Why?'

She lifted her shoulders. 'I'm used to it.' Her expression was ruefully humorous. She had about her a wry young charm.

'Take it calmly. Just hold to your peace . . .'

'Once I thought I had it here. Now . . .'

He glanced up at her. She waited still beside the table, near the kitchen door. Her small, waif-like face stood out against the crudely-daubed green paint of the door. 'Now?' he said.

She smiled softly, shrugging her shoulders again.

Deane put down his fork. He said after a while, 'Stinking drunk yesterday afternoon, wasn't I?'

'Pretty bad. I have seen much worse . . .'

There was a shout from the kitchen behind. She turned her head, then looked back at Deane.

He said, 'Remember what I asked you?'

She nodded.

'So you wouldn't come with me, eh?'

'You were drunk. Ideas change when one is sober.'

Deane said at last, 'Sometimes.'

The cook shouted again from the kitchen. The girl said, 'No peace indeed . . .' She turned swiftly and went through the green door.

Deane sat back. He ate his steak; just as he was finishing Kendrick came into the dining-room.

172

They met one another carefully. Deane said, 'Enjoy your breakfast.' Then he went outside.

He found the old aborigine Albert sitting in the sun. Deane stooped down. 'That feller Jack—him better. Him all right, bimeby.'

Albert stirred within the depths of his swaddling, oversize jacket. His dark eyes peered over the greasy, turned-up collar. 'You see 'im, boss?'

'Yes.'

The wrinkled face considered.

Deane gestured to the camp. 'Bush-fellers, they arrived, eh?'

Albert nodded.

'You belong same tribe?'

He nodded again. 'Dat feller Thomas, him too. Dese bush-fellers, dey havem big meeting, boss. Dey talk what dey do, longa dat bus'ness.'

'What?'

'Tribe, dey punish. Black man's way. Dey smellem out, finish. Dey not trust white mans.'

Deane stood up sharply. Instinct momentarily swung his sympathies. 'They'd better not try it.'

Albert did not comment.

Deane gazed once more at the distant camp where the naked aborigines were gathered. He understood at last that fear of whites for black which could join them . . .

He said, 'All right, Albert.' Then he went back to the police station.

Lawrence received him cautiously. 'Well?'

Deane said, 'What are those damned abos out there like?'

'How d'you mean?'

'I'm a stranger round here. Are they dangerous?'

Lawrence laughed. He stood in the office, police posters fixed on the wall behind him, his square face strong and capable. He belted up the uniform jacket, reached for his cap. 'Getting scared, pommy?'

'To hell I am.'

Lawrence said, 'You're in a wild land. A man's land.' He was derisive. Then his manner changed. 'Are they dangerous? As dangerous as most other collections of humanity. No more, no less. Easy to get along with the best part of the time, stubborn and cruel if they think they've got a grievance ... Why?' His grey, bushed eyebrows lifted.

'I heard the story that they may go looking for the killers of Thomas Clancy themselves. If you don't handle the job.' Deane's voice was harsh.

Lawrence flared up. He slammed his cap back on the bench. 'I warned you this morning. I—'

'Come off it. This time I'm not attacking you.' Deane broke in tightly. 'I want to find out ...'

The copper calmed himself. He snapped, 'Their community has the habit like all others of punishing its transgressors. They won't try anything in my area.'

Deane said, 'You'd better get after Malone first and make sure of that.'

Lawrence peered through the window, over towards the bush camp. He pulled himself round on Deane. 'I thought you'd found yourself in enough of a mess yesterday to keep away from more.'

Deane walked past him, over to the Austin truck.

He told himself it was time he went over to Clancy Rock Station again. He started up, drove off. He caught his breath as he leaned back against the hot leather of the seat and the wounds stung. He pulled himself forward, crouched over the wheel.

From the other side of the street Mex Kendrick came running, his dark, skinny form racing over the hard earth. He called out. 'Hey, Johnny. I'm coming—'

Deane accelerated. 'Keep out of it.'

Kendrick grabbed for the door handle. He wrenched at it, struggling along beside the truck, scrambled inside. Deane cursed, kept driving.

Kendrick panted, recovering his breath. He turned, and his teeth grinned in the sharp face. 'Thought I'd appreciate your company, amigo.'

'I'm not going anywhere.' Deane set the truck on the road out.

'Maybe not. But we'll stick together.'

Deane jammed on the brakes. 'Outside.'

'No, Johnny. Not just yet.'

He lifted his arm. Kendrick jerked forward the flick-knife. His yellow face was cold.

At last Deane started up, drove on. 'You bloody fool. You're only wasting your time . . .'

'Take me for a ride,' Kendrick said.

'I'm going to Clancy Rock.'

'All right by me.' Kendrick crossed his long legs.

Deane chose to ignore him. Kendrick wouldn't be able to interfere. He sat forward, wrenching the wheel over the ruts, conscious of the pain in his back and the

175

ache of his ribs, the contusions on his face and the half-closed eye: he sat still, determined.

The truck plunged over the rough ground. The sun beat down from a cruel sky. Kendrick began to stir, undoing his collar, fanning his face. The burning drought gripped the bush, and only the heat waves moved in the stillness.

They drove through to Malone's paddock. The station lay ahead. The dying cattle strewed the ground. Kendrick said uneasily, 'What're you after?'

Deane did not reply.

Before they reached the station, when its roofs were still distant on the horizon, he saw another vehicle cutting across the paddock towards the road to meet them. The trail of red dust was flung up in the air, founting, racing.

Deane watched it come, and he was suddenly cautious. He speeded up: but the other truck was due to cut them off. A Land-Rover; he muttered sharply, disquieted. He was not eager to meet Malone in the emptiness of the paddock. At the station, he thought he knew how to arrange it . . .

The Austin moved faster, hurtling over the uncertain road. Kendrick glanced at him. 'Slow it down.'

Deane's foot plunged the accelerator into the floor. But the Land-Rover reached the track ahead of them, pulling up to block it. Two men were in the vehicle: Malone, and the jackaroo, Gair.

Deane began to brake. The Austin came to a stop a few yards short of the Land-Rover. They sat still for a moment.

Kendrick said, 'You're on your own, Johnny.'

176

Deane pressed the horn button. Gair was at the wheel. He did not move.

Deane shouted. 'Out of the way.'

The jackaroo looked towards him, but he took no notice. Malone got out of the Land-Rover, carrying a rifle. He walked across to the truck and stood near the open window.

He said slowly, 'What are you doing on my land?'

'Driving to the homestead.' Deane kept it brief.

'What for?'

'To see you.'

'Well?'

Deane said, 'This isn't the place to settle it. Let's drive on to the station.'

'This is just the place.' Malone took a step back, began to lift the muzzle of the rifle.

Deane watched him mockingly. Suddenly he saw the look in Malone's eyes, and he knew that he would fire. 'Hell . . .' He sucked in his breath, ducked wildly, thrusting his head down behind the door panel.

Kendrick exclaimed violently. 'Gaw . . .' He too flung himself aside.

Deane heard Gair call out from the Land-Rover. Malone shouted back. Deane lifted his head carefully, shock of danger tightening his stomach. He was dazed by the astonishment of knowing how near he was to sudden death.

Malone had lowered the rifle, was waiting rigidly.

'You're bloody mad.' Deane swore at him. Malone stood fixed, booted feet in the red dust, heavy squat body, his face wild and empty. The high sun burned

177

over the arid, desolate land where animals and vegetation perished and decayed. Deane checked his breath again, and he knew Malone's madness was a thing not of brain-cells but of this time and place. And he was afraid.

He said wildly, 'Pack it.'

Kendrick began scrambling for the other door. He dropped to the earth, came round to the other side of the truck. 'Leave me out of this . . .'

Malone did not speak. He began to lift the rifle again at Deane.

The young jackaroo yelled out from the Land-Rover. He bounded across. 'Richard! For Chrissake, Richard!' His voice shouted high. He grabbed at Malone's arm.

Malone shoved him away. 'Get off—'

Gair shouted agitatedly. 'Hold it, Richard. For Chrissake don't be a fool . . .' He staggered forward.

Then they stood still under the vast burning sky. Malone brushed sweat from his upper lip. His hand went back to clench the stock of the rifle.

'You can't kill him.' Gair's handsome young face was horrified.

Deane swallowed, dry-throated. Fear had sharpened his perception: he was conscious of the brightness of the light, the silence and breathlessness of the paddock, the sweet dying smell on the air. Hairs grew between Malone's eyes; his skin was thick and pocked. All was vivid and yet remote. The rifle which could kill him . . .

Malone swung the barrel at Kendrick. 'What about you?'

Kendrick rubbed perspiring hands together. 'Count me out, friend . . .' His eyes darted.

Malone's gaze came back to Deane. He was silent. Thoughts passed stealthily behind the bone façade of his skull. At any moment he might react.

Deane stared down from the cab. Malone wrenched the door open. 'Get out.'

Deane climbed down. He was too rigid to move. Then from somewhere courage came, and anger with it. 'You bastard . . .'

Malone's eyes glittered, and the moment came near. He brought up the muzzle. Gair choked in dismay. Suddenly he grabbed for Malone's rifle. Malone bellowed and stepped aside, swinging. He jabbed the barrel at Gair's head.

Gair jumped back and his feet slipped in the loose soil. He staggered, swayed to recover balance. Malone pulled himself to the side, breaking away. He twisted round and the rifle muzzle swivelled again towards Deane. His finger began to slide.

Deane flung himself down. The bullet tore into the truck above him; the sound of the rifle shot cracked over the heavy air. Malone's hand wrenched the bolt rapidly. Deane fell upon him, jerking with pain as his back tore open. He grabbed for the rifle. Malone brought up his knee. Deane turned his body and blocked it.

Gair stood apart, irresolute. Kendrick leaned forward and watched acutely. The rifle broke from Malone's hands; the muzzle dropped low, tumbling into the sand and shale. Deane kicked out, trampling, clogging it. He pulled himself back and stamped on the bolt lever, the open breech; he jumped with both feet on the barrel and felt it give. Then he stopped, breathing hard.

179

Malone bent, grabbed the rifle, wrestled with it. He stared at the fouled breech, the distorted barrel. Then he flung the piece viciously at Deane. Deane dodged. The rifle clattered against the side of the truck, falling.

Malone stood still; his broad chest heaved. Gair waited anxiously near him. Malone took a step for Deane, and then he broke off. His face changed with erratic, unpredictable swiftness. He turned and walked towards the Land-Rover. He swung himself into the seat, set it going. He swung the vehicle round on the track and drove away towards the distant roofs of the station, speeding, retreating. Dust lifted behind, obscuring him . . .

Deane leaned against the cab door of the Austin. His head reeled. Then he straightened up. He looked at the dark, taut figure of Kendrick, then at Gair's scared and sulky face. Gair watched the Land-Rover go, apprehensively. He glanced at Deane.

Deane dropped down to search under the Austin. Water was dripping from the holed radiator. He muttered. Then he climbed into the cab and pressed the starter. He said harshly, 'I'm driving back to Galah. Please yourselves.' He engaged gear and set the truck turning.

Kendrick gave a sardonic shrug. He yanked himself up with one hand and relaxed into the bench seat. He pulled out a pack of cigarettes.

Gair stared wildly down the trail towards the homestead, then back at Deane. He hesitated, standing alone on the trail.

Deane reversed, and the hot soil spurted. Fear

swarmed on Gair's face. He rushed forward, pulled himself up to the seat next to Kendrick. He said rapidly, 'I've done with him. Finished.'

Deane's lip curled. 'I haven't forgotten you helped him string up the abo. You don't get off that easy, cocky.'

'If it hadn't been for me he would've killed you.' His voice was desperate and sullen.

'Blast that. You'll get no thanks.'

Gair looked at Deane. He recoiled slightly. At last he said, 'I had nothing to do with the flamin' hanging. Honest, I tried to stop it—Malone and the abo stockman were the ones. That's fair dinkum . . .'

'We'll go and tell that to the copper,' Deane said. He began to smile fiercely, driving towards Galah through the paddock of glazed-eyed cattle, where the smell of death caught him again.

14

HE pulled up the Austin outside the police station. Again he grinned savagely, thinking of Lawrence's face . . . He turned to Gair. 'Jump out. Now's the time . . .'

Gair got down, stood uneasily near the steps. He yanked up his trouser belt, stuck his hands in his pockets, glancing around the square.

Deane went inside. The policeman was not there; he came out, spoke to one of the half-caste children. 'Policeman?'

''Long bush.' The boy pointed into the distance, towards the hills.

Gair said, 'I'll see him some other time.' He began to walk away.

Deane said, 'You hold it.'

Gair kept walking. Kendrick leaned against the truck and watched.

Deane went after him, swung him round. 'We're staying here till he gets home.'

Gair's smooth-skinned young face darkened. 'Are you out of your perishin' mind? He may be away for a week.'

'On Ration Day? He'll be back.'

'I'm going for a drink.' He stood loosely, his hands dropped.

Deane said, 'Get back to the police office.'

'No.' His voice was hard. He turned and began to walk away, tall and long-legged.

Deane sighed. Then grimly he moved after him, jerked him round, jabbed a fist. Gair came at him with both hands, punching. They battered at one another, ducking, grunting, swearing. Gair was fearful and desperate, struggling wildly. Deane fought with steady endurance, fatigued by pain. They used boots, hands, bodies, clashing together.

Gair went down first, bloody-nosed, hands to his stomach. He knelt in the sandy soil, head bent and the black, curly hair hanging over his forehead. Deane stood over him, near to toppling. Only three or four people were on the street to watch. A dog barked excitedly. The high noonday sun baked the air.

He kicked Gair's leg. 'Get up.'

Gair staggered to his feet.

'Let's go back.'

They crossed over to the station. Inside were a couple of chairs. He pushed Gair into one, then closed the door and sank into the other. Their eyes met. Gair turned away, dabbing his bleeding nose with a handkerchief.

They sat still for a long time. Outside everything was silent. Flies buzzed in the office. A red light glowed on the front of the transceiver. Through the window the bush aborigines' camp appeared deserted in the heat.

Deane got up, found a cupful of lukewarm water at the back. He tried the door to the rear portion of the building where the wounded abo lay: it was locked. He went outside, wandered round to the back entrance. One of Lawrence's native trackers was guarding the door.

Deane stared across at the abo camp again, to the wurlies of the township blacks by the creek, over the empty main street to the shacks. All was silent.

He returned to the front office, offered Gair some water. The young man drank it, shook his head, splashed a few drops on his forehead.

Deane said, 'Better?'

He looked away without replying.

Deane said, 'You can talk to me while you're waiting to talk to Lawrence.'

Gair pressed his lips together.

Deane went over, jerked up his chin. 'You'd better.'

Gair watched him with careful resentment.

Deane said, 'So it's your story Malone and his stockman hung up the aborigine?'

Gair hesitated. At last he gave in. 'Yes.'

'Who shot the other abo?'

Gair spoke in a low voice. 'Malone. He grabbed the rifle from me—'

'And you sat still and kept out of it.'

''Course I kept out of it. I only work for Malone. Nothing to do with me.'

Deane said, 'Where's the stockman today—the black blighter Malone calls Ken?'

'Don't know.'

'Come off it.'

Gair said quickly, 'I tell you I don't know. He's disappeared overnight—probably gone walkabout. Malone was furious about it this morning.'

Deane fingered his torn ear. 'He won't walk far when he shows up again.'

Gair said, 'Who in hell d'you think you are? You're no policeman, you bloody pommy.'

Deane bent towards him. 'I told you last time to stop playing the tough feller . . .'

'You'll know about tough fellers when Malone settles with you.' He stared up, scowling.

'Malone's lost his chance.' Deane said, 'I'll watch you out of here on your way to jail . . . Anyway, you're not so popular with Malone now. Forgotten?'

Gair's face changed as he thought about it. He wiped the fresh blood from his nose.

Deane commented, 'So he's lost you, and he's lost the abo ringer for a while. How many men has he got left to shout for him at the station?'

Gair shrugged.

184

'You'd better,' Deane said again.

At last Gair said, 'He'll be on his own.'

Deane stared at him. 'What about the others?'

'What others?'

'How many does he have working for him?'

Gair eased up. He spoke more freely. 'A few blacks, a handful of boundary riders. Philip Welles who runs the place—but he's soft as a honey ant's belly. Malone's bitch of a wife who hates him . . .' Vengeful satisfaction twisted his face for an instant. 'Malone's crazy, and he's on his own . . .'

'A station that size? He must have more—'

'I tell you that's the lot. Five whites, a half-dozen blacks. And none of them will lift a stinkin' finger to fight for him if he gets in a stew . . . Except maybe Did Billings. He's a squabbling litt'l runtie—he likes a blue for the curse of it.' Gair was spiteful. 'He's a cold-blooded cow . . .'

'Who's Billings?'

'Young chap. Sour-faced joker—wears a green shirt.'

'I've had one go at him before.'

Gair said, 'What are you planning to do?' He cast a speculative look over Deane.

Deane stood up. He was dirty and unshaven, blood-smeared. He wasn't proud of himself, and suddenly now there was no great satisfaction after all in attacking Malone. A bush-crazy cattleman, alone and friendless on his own station . . . He straightened up. He said briefly, 'I told you. Get the copper to pull in Malone. Pin him down for what he's done, and before he causes any more damage . . .' He went over to the window and looked out. After a while he turned round. 'Lawrence is on his way now . . .'

Gair sat forward quickly. They waited.

The truck stopped outside the station. Lawrence got down, a native tracker beside him. His Land-Rover was painted in white along its bonnet, Police. Something lay inside: the children had gathered round to stare in.

Lawrence's face was tight and troubled. He observed Deane's vehicle. He frowned deeper, then he came up the steps, inside the station.

Deane turned to meet him: the copper exploded at the sight. 'Outside! What the hell are you doing in here? . . . Just get out!' He pushed by, worked-up and furious.

Deane said, 'We want to talk to you . . .'

'Get out. I've enough to bother me without you.' He swung his arm. 'Get out, get out . . .'

'Gair wants a word with you too.'

Lawrence's angry gaze flickered over the two of them. Then he snapped, 'I've more important things. I've enough on my plate at the moment. Get out . . .'

Deane said, 'You'd better hear, copper.'

'I told you to go to hell.'

Deane stepped across to him. 'You're going to, Lawrence . . .'

The copper struggled for an instant, on the verge of shouting, unsteady violence. At last he controlled it. He breathed slowly; his face was grey and older.

He nodded to the black tracker who stood in the doorway. 'All right, Harry.' He shut the door, flung his hat on the desk, mopped his brow. He turned to Deane. 'Listen, sport, I'm tired. I'm real tuckered-out. Let it rest.' For once his tone appealed to Deane. His creased eyes were weary.

Deane watched him. 'What's *your* trouble, copper?'

Lawrence paused. Then he turned to Gair. 'Bad news, Ted . . . I've just been out in the bush to pick up your abo ringer's body.' He jerked his head at the truck outside.

Gair was stupefied. 'Dead? Ken dead?' His mouth opened. 'How?'

Lawrence said, 'Somebody stuck a wooden skewer down him—neckbone to stomach.'

Gair gasped. Deane was still. Lawrence unbuckled his belt, hung it over his chair. He sagged without its support.

He said slowly, 'Pretty plain whose trick that was.' He went to look out of the window.

Deane stared over his shoulder, at the bush-camp in the distance, silent in the sun. He shivered.

Gair had paled. Deane caught his eyes. 'Better watch yourself, Ted.'

Lawrence swung round. Gair said, 'Got something to tell you, Edwin.' He swallowed.

Lawrence moved as though to brush him away. Then he waited. At last he said, 'All right. If you must . . .' He sat down slowly on his chair, full of fatigue. 'What is it?'

Gair swallowed again. His fingers played with the braiding of his hat, swung between his hands. Blood had dried on his face, messily. He cast one last look of resentment at Deane. Then he said, 'I was with Malone when he killed that abo Thomas. I was in the Land-Rover with him . . .'

Lawrence's gaze came round to Deane. The expression died into resignation. He stretched out a hand for his pen, heavily. 'You'd better tell me about it, Ted . . .'

Deane sat with his legs crossed, leaning back. He felt no sense of achievement. He felt flat, even low. When it was done, Lawrence pushed across the paper.

'You'd better put your monniker, Ted.'

Gair shuffled with the pen, then he scrawled his name. Lawrence said, 'All right. That's that.'

'What do I do now?' Gair waited anxiously.

'For the moment, what you like.'

'I can't go back to the station.'

'Stay in Galah for a while.'

Gair was still uneasy.

Lawrence said, 'You can sleep here a day or two if you want it.'

'Thanks.' Gair seemed relieved.

Lawrence got up. He unlocked the rear door, went through. Gair and Deane stood silent. In a while he came back. 'The abo's still under. But his pulse and temperature are normal. He's okay.'

Gair said, 'What abo?'

'The one you say you saw shot.' Lawrence was calm now. Gair looked at Deane. Lawrence said, 'Now leave me alone. Just take yourselves out, the pair of you. I've seen enough of you to vomit . . .'

Deane stretched himself thoughtfully. He said, 'Well?'

Lawrence's eyes rose. 'Well what?'

'What about Malone?'

Lawrence turned aside. 'I'll see about that when I've got time.'

188

Deane came forward. 'You can't ignore it any longer, copper. You've got all the evidence now . . .'

Lawrence said, 'Evidence? I've got no floggin' evidence. I've got another statement. From a biased witness who admits he's just had a blue with Richard Malone. It's all on the file—I'll take it further when I can.'

Gair stood back. Deane moved closer to Lawrence. 'You get after him now. You fetch him in—'

Lawrence put on a show of reasonableness. 'Listen. I tell you I'm too damn' busy already . . .' He offered, 'I'll do this for you. I'll call up Malone on the set, put everything to him straight. About the abo, Ted's statement, the lot. And if I don't get a straight reply back that satisfies me, I'll fetch him across and he'll have to explain plenty more from inside the jail . . . Well?' He scrutinized Deane.

Deane said, 'Not good enough. You know bloody well it's not good enough. We've done too much talking about it all round.' He began to get angry. 'You're only wasting time when you know the truth as well as I do. Just drive over and arrest him. Now . . . That's the only way.'

He watched Lawrence flare up: the copper swung round, pulled a pair of handcuffs out of a drawer. 'Just say another word. You say a word . . .' He thrust them at Deane.

Deane's temper smouldered back. He did not move.

Lawrence said, 'Listen, you cow. Get it clear—this is my unlucky day. Too right I'm unlucky. I've got the rations to distribute, I've got that blasted stockman's body on my truck to deal with. I've got to find the

characters who did it and make an example for the rest
. . . I've got you on my neck. My word, you start
bleatin' to me again . . .' He jangled the handcuffs,
trembling.

Deane said, 'Get Malone. He's your first task.'

Lawrence held out the cuffs. 'You've begged for
this . . . Now I've got something else to tell you. This is
going to quieten you fast . . .' He calmed himself, and a
cold smile came to his tight lips. 'Had something through
about you this morning before I started off . . .'

Deane's gaze settled.

Lawrence said, 'Got a reply about you at last. Now I
know why you're travelling through this district. Got
you fixed down at last, and now I'm using the chance.
You're going all the way back to Sydney. You'll run no
damn' further, you'll shout around no more . . .' He
advanced with the handcuffs, face set.

Deane said, 'You won't, you know.' He turned and
burst out of the office, jumping across the steps to the
ground. He walked quickly away. Then he looked
back.

But the copper was not coming after him, and the
doorway remained empty. Once more he began to grin,
with harsh amusement.

He went across the street. The fat woman was standing under the balcony of her store. He leaned against
the post, and they watched the small crowd around the
truck where the stockman's body lay. The policeman
came out at last, scattering them. He drove the Land-
Rover round to the back of the station.

Deane said quietly, 'Heard the news, Margie?'

190

She nodded.

'Maybe the blacks will be after Malone himself, next.'
She didn't reply.

He said, 'If not I'll see the copper gets on to him. The jackeroo's told him all he knows.'

She turned at last, slumping slightly in the loose floral dress. 'Content, Johnny?'

He paused. 'No,' he said. 'Just one of those things. Just the way it goes. I'd have been glad enough to keep out of the whole lousy affair . . .'

She gazed over the square, shading her eyes with a plump hand.

Deane said, 'Malone's going to fall, Margie. One way or the other. So what comes?'

'He's not fallen yet.'

'It's due . . . So?'

'You tell me,' she said.

'The station will go on. Things will keep running, even under somebody else. It won't make any difference to this district, it won't alter matters one way or another. So why should the townspeople care?'

She shuffled her feet. 'Perhaps it'll be the same. Perhaps not. But when you start something tumblin', you never know what else is going to fall. That's why.'

Deane said, 'Life will go on the same.'

'Perhaps. We'll see.' She turned and went back into the store, began dusting down.

He followed her, standing in the cooler shade. 'Anyway, you'll feel better, Margie? You know this town's got something on its conscience—you'll feel better to get it clean?'

'Everybody's got something on their blinkin' conscience. Ain't you?' She said, 'I'm no new-born babe, pommy. I don't expect ever to get me clean and pure . . .' She laughed briefly, cheeks wobbling, then silent.

Deane stroked his finger along the counter. 'Okay, Margie.' He pulled himself up, strolled out.

She called abruptly after him. 'How's the back?'

'All right.' He did not check his steps.

She said, 'Don't mind me anyway—I'm a gutless old cow at times. You act for yourself.' Something in her voice was different.

He turned to look at her; then he raised a hand, walked away.

Kendrick was leaning on the veranda rail, staring at the bushmen's camp. He looked Deane over. 'Still raising the dust, amigo? You'll make yourself real unpopular around here.'

'I'm used to that.'

'You knocked the kid about pretty well. You're better in a blue than I thought.' His thin, hangdog face was sardonic.

'Why don't you whistle for an airplane and get yourself out of here?' He was growing grimly fed up with Kendrick.

Kendrick said, 'Maybe I shall. Can't admit I like the looks of those savages across the way.'

'They're in an angry mood. Don't shift away from the hotel or you'll risk ending up like Malone's stockman —with a spike through your lungs.' Deane walked by, smiling.

He had missed his dinner a long time back; he called

192

in at the café. Inside, the post-office man was talking to the Greek proprietor.

Deane sat at one of the cubicles and munched a couple of sandwiches. The other two men had stopped talking. Then the post-office keeper said, 'How's it going, pommy?'

Deane said briefly, 'Not so bad.'

'Taking it rough these last coupla days, eh?'

'Doesn't worry me.'

The flat-chested, toothless post-office man gave him a quick and gappy grin. Deane stared at the man with a casualness which masked his surprise. The mailman leaned against the counter and began talking again to the Greek. Deane finished, stood up to go.

As he crossed to the door, the mailman said, 'What d'you think about Malone's ringer, eh?'

Deane paused, shrugged. 'He asked for it.'

The post-office man nodded. 'I reckon there'll be more trouble from these bush-fellers.' He jerked his head. 'What's your opinion?' He seemed eager at once to be friendly.

Deane shrugged. 'Don't know them well enough.'

'They get themselves excited . . .' He pushed back his floppy-brim hat. 'Malone's been pushin' it a bit too far, eh? There's a limit . . .'

Deane said shortly, 'You support him, don't you? You live off him?'

The post-office man hesitated. He took it without rancour. 'Maybe we've done that too flamin' long.'

'Why?'

He hesitated again. At last he said slowly, 'Well, I suppose we've all got to live together. Blacks and all . . .'

Deane moved on, outside. The small Greek nodded at him as he went, and smiled.

He walked down the street; he passed by a lounger near the pub, a cattle-hand in tight trousers and sloppy hat, one leg stretched in front of him as he sat. The hand gave Deane a wink of acquaintance.

Deane went on; he was puzzled, not expecting this new sensation of acceptance where before he had met only hostility.

He sat on the veranda during the hot sun of the afternoon, waiting for something to move. Waiting for Malone, or waiting for the policeman, waiting for Jack Clancy to recover and add his words to the file on the murder; waiting even for the bush abos to start something. He was prepared to let the others act.

Heat haze glittered over the afternoon. Lawrence's truck drove out, loaded with dresses and men's trousers, flour and tins of milk, tea. He set up a table near the abo encampment and began to issue the goods, aided by his tracker. The naked aborigines clustered around him. The distant figures moved through the haze, swarming. The bush was flat and still, and the tops of the mountains swayed, far-off.

When the distribution was finished, Lawrence got up and began to walk through the camping-grounds, past the fires; apparently talking to the aborigines, questing, investigating. Deane watched, sitting back relaxed.

The hotel bar was open. He went in and got a drink. None of the Clancy Rock riders was there; only a 'roo-shooter passing through, a couple of hands from the Anderson station in the north.

They stood knocking back the grog, exchanging gossip. The hotel-keeper poured another beer for Deane. He treated him familiarly, among the newcomers. 'What's the drum?'

'Nothing special.'

'Heard the radio?'

'No.'

Fisher's face was lighted. 'Weather forecast shows a change. Reckon the end of the drought's on the way . . .'

Deane wiped his face. 'Won't be too soon.'

'It'll save us.' His voice was quick. 'It'll keep us going on . . .'

Deane stared at him curiously. 'I should have thought this weather was good for trade.'

'To hell with the beer.' He said fervently, 'I want to see rain. I want to see this town alive . . .'

Deane nodded.

'It'll be the end of a nightmare.' He looked at Deane with earnest eyes in the tough lumpy face.

Suddenly Deane was moved.

He went to the door, trying to imagine the town without the aridity and cruelty of the drought upon it, with water in the soaks and flowering on the plain. Without the crazed eyes of the cattle and the drought-desperation of men. When the people could see life in their town and not fear wildly under its death.

He sat in the shade of the hot roof again, staring over the town. It was familiar to him now, the sunlight lying over the dusty street, the feel in the air. The shacks and the stores and the townsfolk, clear-cut in the light, and then the haze beyond into nowhere, into the wildness of the desert and bush. Galah Creek: and now despite all, he found he did not hate it as he had done a few days ago.

The hotel-keeper behind him said, 'Look!' The bare arm pointed.

Deane looked, and Lawrence was coming back from the abo encampment. He was walking, leading the way. Behind him came a couple of naked aborigines, heads downcast, walking dejectedly. Accompanying them were the two native trackers. The remainder of the tribe stood still and silent behind, watching them go.

The small party came across the bare land, towards the police station, caught by the falling sunlight.

The hotel-keeper whistled. 'He's gott'em. He's picked up the bastards.'

The group came nearer. Deane said, 'They killed the stockman?'

'You bet. He's got some guts.'

Deane nodded. Lawrence came first, over to the back of the station, to the lock-up at the rear. He stood at the door, motioning the others inside. The sheepish, un-armed bush-abos with their pipey black shanks and fierce faces stepped in. Lawrence followed, closing the door.

The hotel-keeper said, 'He's fast. Took some doing, among that bunch.'

'Surely did.' Deane said, 'Pity he didn't go after Malone so quickly.'

The hotel-keeper glanced at him. Deane put down the beer-glass, walked over the street.

Gair was in the front office of the police station, Jack Clancy lay in the rear room; Lawrence and the two bush killers were in the jail section.

196

Deane gazed across the land at the abo camp. Mex Kendrick came like a shadow from nowhere and stood at his side. 'Slick job, amigo.'

'Yes.'

'Now what?'

Deane shrugged. The other inhabitants of the town were there too, looking towards the crude encampment from where the murderers had come. And the naked black aborigines stood on the edge of their camping-ground and stared back at them, silently.

15

EVENING came down, darkness blowing over from the bush. Fires burned in the aborigines' camp; smoke swirled, white and thick beyond the eucalyptus, and shadowy figures passed across the lights.

The town was quiet and aware. Over the hotel veranda the dim lamp gleamed; the square was silent, and long dark patches of shade filled the corners of the street. Above, the stars glowed. Nothing moved.

The fires in the camp burned down. The dry dead arms of the mulga trees swayed, creaking in a slow wind. Some time during the night Deane heard a noise, a crack of sound which broke across his half-sleeping brain. He threw himself from the bunk, expectant of anything, then went to the window.

He heard the thresh of an engine cranking, starting.

A vehicle drove away, somewhere across the dark, off into the night. All was quiet again.

He peered out, standing on the veranda. He could see nothing: only the white starlight in the middle of the street. He went back into the room, to the jerky rattling of Kendrick's breathing opposite him.

Daylight came. Deane stirred, began to dress. He moved painfully, still stiff from bruises and lacerations. Across the street beyond came a shout, the skeltering of feet. Then silence, then more voices, then sharp silence again.

He stepped through to the veranda. A group of men had gathered across the square, by the rear of the station. The fixity and intensity of their poses, caught still, struck through to him: the bunch of men was staring at a pile on the ground.

He belted his trousers, swung himself over the rail and strode swiftly to see what went on.

The morning light was fresh and hard. A half-dozen abos from the township settlement huddled together, gazing at the body which lay heaped against one wall of the police station. On the other side of the group were the whites, unspeaking, sober-faced; the post-office man and Fisher, the hotel-keeper, the travelling 'roo-shooter. The Greek from the café bustled to join in. Between them was the body of the aborigine, Jack Clancy; the front of his skull was blown away and the blood was caked dry.

The policeman came along, hurrying, his face twisted with harsh despair. He pushed through, knelt by the body. Then he looked up.

One of the aborigines muttered something. They stood sullenly bunched, hostile.

Lawrence pulled himself to his feet. His gaze moved around the circle. Deane burst out. 'God damn you. You let—'

'No.' Lawrence shouted wildly, waving him down. 'Not now—' He called out for his trackers. He swung on the whites again, aghast, choking with emotion. 'Ted Gair's still in the front. The bush blacks are locked in the jail. Why in all cursed, unholy hell—'

The hotel-keeper caught at his arm. 'You've gotter do something pretty fast, Edwin. Time to move now ...'

The abos stood still, unwinking gazes fixed on Lawrence. He pointed quickly. 'You boys, you go 'long back. Them fellers done this, me catch 'em. You see.' He stood half-dressed, his uniform unbuttoned, his boots unlaced. He was a desperate, dishevelled figure, white-faced.

The aborigines turned slowly, went back to their camp. The Greek chattered under his breath. 'Trouble. More trouble, when they go tell their friends across the way ...'

Lawrence poked at the broken window behind the body, staring inside, tearing his hands on the jagged edge. 'This way ... Why to Christ didn't I hear them ... ?' Again he stared impotently around their faces.

Deane said, 'Someone drove out of town during the night. I was awake, but I couldn't see anything. *You* were supposed to be guarding—' His gaze attacked Lawrence.

One of the trackers called out. Lawrence followed him

rapidly across the square. The native examined an invisible mark in the road, pointed over the back of the shacks, out on the road towards Clancy Rock. His finger aimed at the horizon . . .

Lawrence's voice was anxious. 'What was it, Harry?'

'Dat feller Land-Rover.' The black tracker said, 'Long' last night, boss. Fif', six hours maybe.'

The men stared at one another. Deane swore again. 'Copper, you've done it now. You took the responsibility—it's yours.' His fists tightened.

'All right. All right, then.' Lawrence said unsteadily, 'I'll take that responsibility.' He gazed along the track.

The hotel-keeper gripped Deane's arm. 'Give him a chance.'

'He's had his damned chances.' Deane looked down at the flung body. The bandages were still wrapped round the naked stomach. The blame lay on himself as much as on Lawrence. He had brought in the poor blighter . . .

His stomach turned and he was bitter. 'Why didn't you watch him?'

'I never thought that—' The copper stopped.

Kendrick came walking fast over the street to see what was happening. The black button eyes fastened on the body in the dust. His hand moved to his pocket. Deane saw it grip reassuringly on the butt of the .45 automatic. 'This gets nasty.' The thin face was queasy. Kendrick was a city man: he did his dirty work on city sidewalks, under shaded bar lights. Violence in the outback was elemental . . .

Deane swung back on Lawrence. 'You promised to look after him. You're as guilty as Malone—'

'I'm going after him.' Lawrence's mouth shut firm and his jaw set. 'There's no other way.' He made his decision; his face showed a fierce, grim resignation.

'You left it late.'

The post-office man commented, 'How'd he know about the abo?' They looked at one another.

Lawrence stood still. 'I spoke to him yesterday over the transceiver. Told him I was holding the black here, and why. I asked him to come in and explain himself . . .' The hard sunlight fell over his desperate face.

Kendrick grinned cynically. 'He came, all right.'

'He's a killer.' Lawrence said flatly at last, 'He's a killer. There's nothing else to do.'

Margie Thompson waddled across from the store. She looked down, and her hand rose to her mouth. The abo's body had not been moved. Soon the flies would swarm round. Grace Fisher was watching from a window of the hotel. Deane's stare passed over the street. Everyone was there, everyone was looking. From the outer edges of the town as well the aborigines were looking, gazing at the murdered body of their tribesman, at the whites who stood in a close group and talked . . .

The sun was striking hotter. Lawrence's face was a fixed, self-accusatory mask.

Deane said, 'He must have had help.'

'I'll find that out. I'll guess for Did Billings . . . I'll settle this.' His bitter gaze moved round the group once more; then he turned, hurried to the office.

They waited. In a moment he came out with hat and belt, revolver holster. He called to one of his trackers, crossed to his truck.

Deane started after him. 'I'm coming.'

Lawrence snapped at him. 'No, you're not—this is my business. Keep out. Just keep out the way I've told you a dozen times before . . . Just see where your interferin' has got us.' He was violent with tension.

'I'm concerned too. I want to see what happens.'

Lawrence shouted, 'Stay away, Deane. Just stay away.' He jumped into the police truck, set it moving, bumping forward.

Deane ran to the Austin. He reversed, pulled her round.

Kendrick's lean figure darted over the square towards him. He grabbed, fell into the cab beside Deane.

Deane braked, shifted gear. 'Get out of it.'

'I'm sticking tightly to you. Eager to get away from this place—gives me the willies. First time I ever chose to be where the law is . . .'

'Get out.'

'I'm staying with you, amigo. Enjoy it.' A brief grin showed the pointed teeth.

'Then sit tight and take it.' He jerked the wheel, slamming forward. Kendrick clung to the front panel.

'Hold it, pommy.' The hotel-keeper was rushing along the side of the truck. 'Wait for us—' He panted.

Deane touched the brake. 'You too?'

'I'm coming.' Fisher reached for the back supports.

'Speed it up.'

He scrambled over the side, tumbling into the rear compartment. He crouched under the roof behind Deane, bare-shouldered in his dirty singlet, face concentrated in suspense. 'I want to be there—this means plenty to us.'

Deane stabbed down the accelerator again and the truck plunged forward into the dust-cloud raised by Lawrence's vehicle. He stared at the hotel as they rocked by. Elsa stood on the steps, gazing out intensely across the road. They looked at each other and then the Austin swept by.

Deane drove fast, racing. The new springs flexed and bent as the vehicle crashed over the rough road. They caught up with the police truck several miles on, in the wastes of the bush. Lawrence pulled suddenly to a halt in front of them.

Deane stopped the Austin. Lawrence got out of the Land-Rover, walked the few yards back across the track, stood near the cab window.

He was sombre and grim, the parched mulga scrub behind him. He said sharply, 'I've got enough to think about. Just give me a flamin' chance. Turn around and drive back, sport.'

Deane said, 'We're all interested.'

'Turn around.'

He shook his head.

The hotel-keeper said, 'We'll stay on the edge of it, Edwin. Trust us.'

'Get moving.'

'No.'

Lawrence turned away. He went back to the Land-Rover and drove on.

Deane came behind, plunging past the tinder-dry

spinifex under the glare of sun. Occasionally he stared at the water-gauge on the panel: the radiator was still holed from Malone's bullet. A couple of miles inside Malone's territory the police Land-Rover ran to a quiet, dying halt.

Lawrence stared furiously behind. Then he got out, lifted the bonnet. Deane stopped, went up to him. 'Want a hand?'

The look on Lawrence's face blistered him. He and Fisher gathered round, gazing at the dead engine. The resentful anger swept over Lawrence. He jiggered unsuccessfully with the carburettor. The starter turned the engine over, but it would not fire.

Deane said, 'I suggest—'

Lawrence snarled, 'When I want your perishin' suggestions I'll ask for them.' He bent over the open bonnet again.

Fifteen minutes passed. The heat of the bush grew heavier. Sweat poured off their bodies, soaking. Deane said at last, 'Check from your blasted coil.'

The fault was in the main lead. Lawrence refixed it. The engine swung, kicked into life.

He went back to his seat. They drove on, the Austin still following the Land-Rover. It was mid-morning when the roofs of the homestead came into sight.

The air baked; dying cattle watched apathetically at the roadside, the scabrous hides clinging over their razor-sharp spines. Kites wheeled in a high sky. Farther on, Malone's station-hands had been burning great heaps of the corpses. A pile of smouldering ash and bone smoked in a cleared area of the scrub; the smell of burning still

lingered sweet, horns and skulls among the embers . . .
The trucks ploughed by.

Inside the Austin Kendrick sat motionless; Deane and
the hotel-keeper were untalkative; each was obsessed by
his own thoughts, oppressed by the external scene of
desolation. Deane watched the symbols of death as he
drove, the circling birds and the skulls, the silence and
emptiness of the bush; the rage he felt towards Malone
evaporated, scorching away in the pitiless bareness of the
bush beneath sun and long drought, its infinity and indifference, the madness it brought.

To see the dying of all that lived . . . He thought slowly
of Nancy Malone among this, and the desperation of a
man who built his station and strove to hold it . . . And
always that terrible fear which must come from this end-
less and ancient land when there was drought, when time
stood still . . .

The homestead was nearer. The tracker in Lawrence's
vehicle ahead flung out a black arm, pointing towards
the horizon, towards the far hills. He chattered to
Lawrence. Abruptly the Land-Rover pulled up.

Deane stepped on his brakes. He and the other two
climbed out again, walked forward to the police truck.
The tracker was yammering away, his arms gesticu-
lating.

Lawrence stepped out of the Land-Rover, shaded his
hand against his eyes. Nothing was in sight.

Fisher said anxiously, 'What's wrong?' He rubbed
his hand against his vest, scratching his chest, eyeing
Lawrence.

Lawrence said, 'Bush fire.'

Kendrick's apprehensive face tightened. He wrinkled up his eyes, staring. 'Nothing to see—'

'You'll find out.' Lawrence fingered his chin in uncertainty. 'Travelling across . . .'

Deane said, 'Bad one?'

'They're all bad.'

The tracker chattered excitedly again. The policeman said tightly, 'Heading for the station. He reckons the bush blacks have started it.' The angry violence was gone from his face. Now he was cold and full of foreboding, staring out.

They stood in a small uneasy group upon the plain. A high sun flared in a pale sky, over the empty quiet land where only a faint wind stirred. The ground was shadowless. There was a brooding silence over the bush, as before thunder.

The hotel-keeper twitched his beaky nose. 'Smoke! Hell, I can smell it!'

Lawrence said, 'Don't get blasted jumpy. That's from the pyre way back—' He jerked his head.

The tracker turned to him again. 'Plenty big fire, comin' fast, boss. Comin' plenty too quick.'

Lawrence moved sharply. 'Let's go . . .' He flung himself into the seat.

They ran back to the Austin. The two vehicles thrust forward again, over the paddock.

Kendrick said, 'We'd do best to get out fast. I mean fast.' He looked nervously towards the distance.

'You wanted the ride,' Deane said. 'Maybe in a half-hour you'll be burning like the bloody bullocks . . .' He grinned.

Fisher clung to the side of the truck. He licked his dry lips. 'Wait till you see a proper bush fire, pommy. You won't make jokes . . .'

Kendrick shifted on the seat. Nothing in sight . . . He said hopefully, 'Anyway—maybe that tracker's imagining it. How in hell could the blacks have started one? The town's miles away—they couldn't get here as fast as a car.'

'Them blacks do some sweet funny things.' The hotel-keeper watched the sky ahead. 'You'd be too darn surprised what they can do.'

Kendrick stared back at him. They covered the last mile, reached the homestead. The truck raced into the yard. Nobody was about. All was quiet. For a moment the calm made an anticlimax after their urgency, after their concentration. Malone's pumps still supplied water to spray the yellowed lawns. The wilting flowers kept a remnant of bloom.

Deane gazed between the buildings to the distant horizon. And then a flash of fear went through him: now at last he could see the fire. Now it came home. A long, low line which moved over the bush, away in the hazed and smoky distance, remote like the dust of a vast army advancing. A stir, a breath of disaster. He felt something catch in his throat.

Lawrence jabbed his finger on the horn button, kept it there blaring. The calm broke into confusion.

An abo servant girl came tumbling out of a door. She screamed and began to run about, flapping her arms like a chicken. Suddenly the homestead burst wide open into tumult.

The other lubras dashed out of the kitchen. The manager, Welles, hurried on the veranda, followed the gaze of the frightened aborigines. His nervous hesitation changed; he was galvanized into activity, racing down the corridor.

Nancy Malone came outside. She gazed at Deane. She was cool and still on the veranda; for a moment he looked back at her, remembering how he had first seen her there, ripe and violable, the pulsing of desire. Then urgency revitalized him. He said, 'Grab what you want. Step on it, fast.'

He stared behind her, into the room with the china and silver. Then he ran back down to Lawrence. Welles stood beside him, gazing wildly at the frontal wave of the fire, which was sweeping closer, rising up in the sky with great gouts of flame.

Welles choked, 'It'll get the buildings. It'll finish us.'

'Nothing we can do.' Lawrence grabbed him. 'Who else is to come?'

Welles stared around, at the frightened abos, the two white cattle-hands. Nancy Malone came steadily down from the veranda, carrying a zip bag. The green-shirted cowboy named Billings rushed across from the office, face urgent.

Lawrence seized him, shook him violently. 'I'll see to you at the station, Did.' Then he flung him back. 'Get your car. Drive some of the abos back to Galah. Now get moving.' He turned on Welles. 'Where's Malone?'

'I don't know,' Welles said. 'He was with Did.'

208

Lawrence turned again. Billings muttered, 'He's still in the office.'

Lawrence stared at the building. Then he snapped at Welles, 'Who's left?'

Welles' quick eyes scanned the yard again. 'That's all. Two of the riders are out west—miles away. They'll miss it.'

'Dead sure?'

'Yes.' His high voice grew unsteady as the fire swept closer.

Lawrence yelled, 'Then start going—' He glanced at the hut from which Billings had come. Then he ran towards it.

Deane watched him go. He raced after the policeman. Now the sunlight was dimmed over the yard; smoke was rising to the sky, obscuring the light. A heavy noise came from the distance over the paddock, breaking in a wave, crackling. In the air was a feeling of doom, of impending catastrophe.

Lawrence stopped at the door. 'Keep out of this.'

'Don't waste time.' Deane was breathing fast.

'Get out, pommy.' Lawrence stood with his back to the door.

'Come on.' Deane thrust by him, wrenched at the handle. Lawrence jerked round, threw his own weight on the door. It crashed open, and they ran inside.

Malone was standing at the window, his back turned, watching the fire rush down.

16

LAWRENCE recovered himself. Then he said urgently, 'Time to go, Richard.'

Malone turned slowly. His heavy face was dazed and stricken. He said, 'Nothing to do. Nothing . . .'

'It's too late.' Lawrence's voice was bare of emotion.

'Drought. Now fire . . .' The big head sagged. Malone was too shocked to know fear. The hard grey eyes were empty as a dead man's. He pulled himself round to the window again.

Lawrence said, 'Come along, Richard.'

Malone did not move. Lawrence eyed Deane, jerked his head.

'About time.' Deane was getting worked up. The roar of the fire came nearer. It was a red, leaping monster through the glass, tearing across the bush with many heads and greedy hands.

They seized Malone. He let them lead him quietly across the office, out of the door. The smell of burning on the slow wind caught them in the yard. Then he began to struggle, cursing.

They got him to the Austin truck, arms in a lock, thrust him inside. They held him while the hotel-keeper drove furiously out.

Kendrick crouched low on the front seat, gazing at the writhing face of Malone, at the red glow in the sky which came closer. 'Speed it, speed it . . . Keep her going.' He panicked, staring at the tall, twisting forest of flame.

The other vehicles had left the station. The truck crashed on down the road, past the starved scrub which would ignite like powder, past the dead trees, wildly fleeing. Now the sky had turned dark as night beneath the blowing smoke, and the noise of burning hung under the smoke and was an angry, shattering roar. The cattle in the paddock knew too well what was to come: terror glittered as the last emotion into their glazed eyes, flickering under the apathy. The gaunt, bony shapes tried to move, to stagger a few futile yards: then they stopped, heads turned away, waiting for death like stoical, pitiful beings that have endured too much . . .

One bullock limped to the centre of the road, then collapsed. The hotel-keeper stood up in the seat and stamped his full weight on the brake. The Austin burned to a halt with smoking rubber.

Malone flung himself to the side, knocking Deane away. He broke from their grip, fell to the ground. He pulled himself up, began running back down the trail towards the homstead a half-mile away. He ran unsteadily, swaying, shouting out inaudibly, a lumbering, terrifying figure in his madness. The smart Ashburton hat fell off, the well-cut jacket flapped open behind him. He ran wildly, wildly on, through smoke and death and red fury of heat and flame . . .

Deane jumped after him. He stopped, yelled to Lawrence. 'Going back?'

'No!' Lawrence shook his head swiftly. 'If we do we're all done—' His face was smoky and terrible. 'For God's sake get in! Save ourselves—' He banged Fisher on the shoulder. The Austin sprang forward, off the track, past the dead bullock. For a moment the wheels spun in the loose dust, sliding . . . Deane shoved, panted, scrambling. Then the drive picked up. He pulled himself on board. They climbed back to the road, plunged forward again.

Deane stared behind. The staggering figure of Malone was far down the track, still running on, heading into the dark of the smoke-cloud, to the doomed homestead.

Deane wiped his face, trembling. Lawrence's expression was sick and awful. He said, 'Once he was my cobber.' He covered his eyes. The Austin raced on, parallel with the fire, breaking away. Heat and red flame burned in waves upon them. Touch and go, life or death. The mulga trees exploded with crashes like the firing of great guns, shattering, flaring as huge torches.

They skidded across a stony patch. Ahead was the edge of the paddock, the boundary-wire and fencing. Suddenly they reached the edge of the fire-front. The wall of flame ended, the smoke thinned out. The air seemed swiftly cooler. They were through. The truck drove farther, to the open gate. Beside the wire were drawn up Billings' Chevrolet, the Land-Rovers. The inhabitants of the station waited, abos and white, shivering gins, Welles and Nancy Malone. They stood watching the Austin drive out.

Nobody spoke. Lawrence got down slowly. Deane followed him. They gazed back at the fire.

It raced by, tall and deadly. Roaring, crackling, the exploding of the trees; the air was dry, and the red terror swept over the land, fast as a car could travel, balls of fire driving house-top high above the ground. Smoke and fury and destruction, a black sun in a dark sky, chaos and devastation. It passed over the bush, dying out far-off to the gibber flats; a black waste lay behind it, covered in thick smoking ash. They watched it, standing silent, waiting. The red glow died; even the slow wisps of smoke passed away and the blackness was desolate, cooling.

After a long time Lawrence turned tiredly to Welles. 'Take the others on into Galah—and watch Billings. Notify them down the line about the fire.' He spoke with empty weariness. His expression was obscured behind a coating of ash and dirt.

Welles nodded.

Lawrence said, 'I'm going back to the homestead . . .' He walked slowly to his truck.

Deane hesitated. Then he said, 'Let me come?'

Lawrence shrugged his shoulders without reply.

Kendrick said, 'Not me this time, amigo.' He was still shaky, the city suit scorched and torn, black soot over his sallow skin. 'I'll take the Austin in. I need a drink . . .'

Deane sat beside the policeman in the Land-Rover. They drove back slowly over the way they had recently come. The ground was still hot and the rubber of the tyres stank high. The thick black ash lay everywhere, clogging the wheels. A burnt, ashy smell hung over the landscape. They passed the bodies of cattle, blackened and contorted shapes with twisted limbs. They sat

213

without speaking; then they drove silently into the homestead yard.

The place was a dark ruin. The fierce heat of the fire had charred all to wreckage. Tumbled iron sheeting, piles of debris, roofless walls. The yard was littered with burned metal and wood; the same silence hung everywhere under a white, glaring sun.

They stood in the stillness, stirring the fine ash with their boots. Embers still smouldered. Deane said at last, 'Wonder how far he got?'

Lawrence stared around, shrugged his shoulders once more. He wandered off alone across the yard.

Then he looked back at Deane. Something in his face made Deane chill. He walked to the spot. There was an iron water-trough in the yard; most of the water had gone, and the metal was still hot. Malone's body lay inside it, boiled like a lobster.

Lawrence said harshly, 'He got here.'

They drove back to Galah. It was past noon, and the day was stifling. The heavy air lay thick on the ground. Dead wind fanned their faces as they drove. The police truck stopped in the square.

Deane got out. The other vehicles were drawn up in front of the hotel. There was no shade. The scattered leaves of the eucalyptus hung low; beneath the bare trees, the township was alert with fear and horror. The children and the street-abos watched the hotel. Old Albert leaned upright, his melon-shaped head pushed forward, face screwed up in attention. The long street lay under the sun, away into nowhere. Light washed off the iron roofs, gleaming against the black walls.

Deane turned to the policeman. 'Going to arrest the blacks for firing it?'

Lawrence shook his head. 'Who's to prove?' He said flatly, 'They'll think they were justified anyway . . .' He stood in the square, under the burn of the sun. 'That's the end of it. It's all done, now . . .' He walked over to his shack, holding his shoulders back, holding himself upright.

Deane went to the hotel. It was crowded with the refugees from Clancy Rock: Elsa and Grace Fisher were handing out cups of hot tea. Among the cluster of people in the hallway he saw Nancy Malone. He could not face meeting her and he went through to his room. He sat alone on his bunk, closed his eyes. He felt still the smoke and desolation of the dead land . . .

Late in the afternoon Lawrence came over. They were assembled on the veranda; Welles and the white hands, Nancy Malone. Lawrence had cleaned the black grime from his face; he was sombre and curt. He said, 'I've arranged for an emergency plane to come in this evening with help. I'm putting Gair and Billings on it— they'll be better off in a different town.' He gazed quietly at Nancy Malone. 'If you feel you'd like to get away a few miles to recover, Nancy—'

She said, 'I would. Oh, I would. I'll take the plane out.' Then she dropped her head.

Lawrence nodded. He left them, walking deadbeat. Deane stared across the square to the bush aborigines' encampment; but it was gone. They had drawn their rations; the killing of their tribesmen had been avenged. They were returned to the bush.

The plane was due in at six. Lawrence came to pick up Nancy Malone and take her to the airstrip. Billings and Gair sat stilly in the truck. Malone's widow came to the doorway. She saw Deane on the veranda.

She waited for an instant. Deane crossed over to her and they stood in the shade of the doorway. At last he said, 'You'll find something better, Nancy.'

She nodded.

'Look after yourself.'

'Yes.'

Below through the doorway he could see the sunlight and the Land-Rover in the patch of light. Her face was hard and yet vulnerable.

Suddenly she said, 'I'll have the station rebuilt. I'll do that for him. I'll see it goes on . . .' He heard the strain and desperation in her voice.

He nodded.

'But I'll never come back,' she said. She went down the steps, climbed into the Land-Rover.

Kendrick rushed out abruptly, calling down from the rail. 'I'm joining you.'

Lawrence's head lifted. He paused. 'Just about room.'

'Hold on.' Kendrick fetched his bag. He came back to the veranda. His dark eyes slid over Deane. 'So long, Johnny.'

'Going?'

'Smart guess.'

Deane said, 'Had enough?'

Kendrick's face twisted. 'Too right, pommy. I'm a city feller, damn me I am . . .'

'I can keep the money, heh?'

The mocking gaze raked him contemptuously. Kendrick said, 'You've got no flickin' money. You're not bloody smart enough . . .' Then he pulled down his hat, paced to the truck. 'So long, amigo.'

He crammed into the back of the Land-Rover beside the tracker. Lawrence drove off.

Deane stood on the veranda and watched until the truck was out of sight.

A few minutes later the transport aircraft came in from the hills, wings shining, settled down in the distance where the airstrip lay.

Deane sat there longer, silent in the sun, unmoving. Later yet, he was still sitting in the same place when the aircraft rose up once more, turned for the south, flew away . . . The sun began to set.

At last he went to eat something; he walked into the dining-room for a meal. Welles and the other hands were finishing. Deane sat alone at the table after they had gone. He drank tea; then he turned to the waitress.

She watched him quietly. He said, 'Busy day, Elsa.'

'Very terrible . . . And I've missed it all. I feel I should have done something.' She had a brave, determined personality behind the small face and the luminous eyes.

'You handed out the tea.'

She exclaimed impatiently. 'I think of the cattle. A cruel end . . .' She was grave.

'So was Malone's.'

'He chose it.' And now she was brief.

Deane pushed back his chair. 'It's the end, surely enough.'

The girl waited near him, holding the cleared plates in her hands. 'The town will be quiet. Maybe it will die, and I shall have to go.'

Deane shook his head.

She said, 'Some have gone already. And you?'

'I'd better follow. They won't appreciate me around here. At the best of times I'm a lousy citizen—I have to keep moving anyway . . . And in this place I've caused plenty too much trouble already.' He turned away.

She said in a rush, '*I* like you, Johnny.'

He glanced back. She was watching him, her young face slightly flushed, reserved and yet frank. Her head was high and she had a tremulous, defiant expression. Deane said, 'Thanks, Elsa.' He went outside.

He walked down to the store and found the fat woman. They sat on the step, looking out at the dying sun going down red in the sky. A touch of coolness lay over the air. She scratched her stomach, sighed. 'Not much to say, pommy. Is there?'

'Not much.'

She stared at the last sun with a plump, quiet face. 'Life goes on, y'know. People can stand up to anything and survive. There's somethin' put in them. That's why I never worry whatever the rest of the bloomin' town thinks. And they'll find it for themselves now . . .'

Deane said, 'Someone else will run the station. It'll all go on for a while. Then things may get worse, they may get better. Just take what comes . . .'

She shrugged. At last she said, 'Poor bloody old Richard Malone. A stinkin' finish for a big man. But it was due . . .'

'Yes.'

'You always get what you ask for.'

'Maybe.' He sat looking over the street, at the red, falling sun over the silent bush, the far sky, and the small group of shacks beneath it. The dusty bottle-tree and the broken buildings, the iron roofs and the wooden walls. It was all still. Suddenly Deane said, 'If I wanted to stay on after this—do you think they'd have me, Margie?'

She jerked her head round to him. 'You? Here?' She stared at him.

Deane looked back down the street. 'Yes.'

'I thought you wanted to get to blazes out.'

'Maybe it grows on me.' Deane spoke roughly, almost angrily. 'I've seen some shocking awful things today. I've been afraid and I've cursed the blasted district to hell. Its inhabitants with it . . . But underneath, something's holding me . . .' He rubbed his forehead, in perplexity. 'I've kept running a long way, too long, too damned long. Somehow I want to dig in now. I'd settle for here, despite the lot.'

He was slow-voiced, not looking at her. He traced his boot in the dust, suddenly hesitant. The Austin was parked across the street, and nothing more to hold him; and he did not want to go, away on the dust-road into nowhere and loneliness.

This land was aridity and roughness, death and devastation: violence of men and cruelty of sky; but it had power and strength besides, and it held him. It was beginning to suit him.

Her eyes searched his face. At last she said, 'Then

stay, pommy. Stay, and welcome to you. You'll find the people round here ain't so rotten lousy. Stay with pleasure, and be one of us . . .'

Deane said, 'You know what I am and what I've been—'

'Ah, to hell with that. We've all done some lousy things, we ain't so clean and pure. You're all right, Johnny.'

He said doubtfully, 'Sure?'

'I'm sure, right enough. You talk to Edwin Lawrence, an' if he's got nothing against you you're fine.' She got to her feet. 'You could find somethin' to do, even in a crumblin' little corner like this. You could make yourself a job—and I'd let you run on tick from the store until you do . . .'

Deane stood up beside her. Suddenly he was humble and deeply moved. He knew that humility was, after all, the hardest penance. He looked at her, untidy bundle of flesh, fat cheeks and blowsy dress, against the red flaming of the sunset, her good, generous face. He said, 'Many thanks, Margie.' Then he stepped down to the street, walked slowly away.

He went back towards the hotel. Lawrence's police-truck swerved past him, back from the airstrip, pulled up at the station.

Deane called out quietly. 'Lawrence.'

The policeman turned on the step, waited for Deane. His face was cautious, showing no expression.

Deane stopped beneath him. He said, 'What are you going to do about me, copper?'

Lawrence hesitated. Then he came down the steps,

stood level with Deane. The faded grey eyes looked him over. Then he gazed away. At last he said, 'Not so proud of what I've done at times lately, sport . . .' He spoke thickly. Then he lifted his head. 'I'm not going to condemn you. Let's just call it straight. . . .' He began to turn away.

Deane said, 'I'm proposing to stay in Galah Creek, Lawrence.'

Lawrence looked at him. Suspicion lighted momentarily. Then he repeated, 'Stay?'

'Yes. Settle here.' Deane's smile flickered briefly. 'I find I like it, after all.'

Lawrence's gaze hardened.

Deane said, 'True enough. I've just been to talk to Margie Thompson. I want to stop—find myself a living.' Suddenly, against his will, a note of anxiety came into his voice. 'What's your answer?'

He stared straight at the policeman. Lawrence was silent. Then he said, 'In this fallin', finished town? You're crazy.'

Deane said, 'It'll last a while yet. I want to stay.'

'Certain?'

'I'm certain.'

Lawrence watched him still. Then some of the bitter load lifted momentarily from him. He relaxed, and warmth softened the craggy face. 'My word, my flamin' word, I never expected that . . . All right, pommy. All right, then.' He was almost eager. 'Just shake.' The square face smiled at last.

'Okay, copper,' Deane said, taking the hand.

He crossed over to the hotel, walking quickly, excited

suddenly and not even sure what he had to be excited about.

He stood on the veranda, gazing out. As he watched, the dark came down swiftly. The air was cool, even damp; the leaves of the bottle-trees and the eucalypts trembled, sighed. It was a plush sky, brilliant.

He saw a red glow in the dark below, a puff of smoke. He said, 'Come up here, Elsa.'

The girl turned her head, walked round the corner of the hotel, stepped on the veranda beside him. From the window of the bar beyond came the sound of noisy voices. Malone was dead and Clancy Rock Station had tumbled; but the rest of life went on.

She leaned against the rail, breathing out the cigarette smoke. 'This lovely evening—after such a day . . .'

'I know.'

She said, 'I still love this town. At all times it's hard work and hard words. But there is something else here, there is something else . . . I'm glad to stay.'

He nodded.

She said, 'You?'

Deane said, 'I've decided to stay as well.'

She turned round abruptly. 'You? But you are travelling on, you are English, you are heading for the north—' The words sprang out of her.

Deane said, 'I'm stopping here. If I can scratch a bare living. Maybe this will make home . . .'

'Truly?'

'Yes.' He smiled.

She relaxed, leaning against the post. Then she too began to smile in the yellow lamplight.

There was a sudden noise on the roof. An abrupt tap, short and quiet. Then again. They stared at one another.

Then silence.

Then the drumming began, echoing on the tin roof, booming, crashing. A wind sprang up, lashing warm rain against their faces.

A roar of exultation went up from the bar. Deane started forward and the girl followed him; they ran down into the square. The men from the bar rushed out and they all stood in the street under the onslaught of rain, faces lifted, drinking it.

Rain came down under the vast stars, and the men were shouting and cheering. Deane stood with the girl.